AF616424

SOCIETY FOR EXPERIMENTAL BIOLOGY

SEMINAR SERIES · 12

BIOLOGY OF THE CHEMOTACTIC RESPONSE

BIOLOGY OF THE CHEMOTACTIC RESPONSE

Edited by

J. M. LACKIE

Lecturer in Cell Biology, University of Glasgow

AND

P. C. WILKINSON

Reader in Bacteriology and Immunology, University of Glasgow

CAMBRIDGE UNIVERSITY PRESS

Cambridge

London New York New Rochelle

Melbourne Sydney

Published by the Press Syndicate of the University of Cambridge
The Pitt Building, Trumpington Street, Cambridge CB2 1RP
32 East 57th Street, New York, NY 10022, USA
296 Beaconsfield Parade, Middle Park, Melbourne 3206, Australia

First published 1981

Printed in USA by
Vail-Ballou Press, Inc.

Library of Congress catalogue card number: 81-6150

British Library Cataloguing in Publication Data

Biology of the chemotactic response. –
Society for Experimental Biology
seminar series; 12)

1. Cells – Motility
I. Lackie, J. M. II. Wilkinson, P. C.
III. Series
574.87'64 QH647

ISBN 0 521 23305 4 hard covers
ISBN 0 521 29897 0 paperback

CONTENTS

List of contributors vii *Preface* ix

Introduction xi
J. M. Lackie and P. C. Wilkinson

Chemotaxis as a form of directed cell behaviour: some theoretical considerations 1
G. A. Dunn

The relationship between leucocyte adhesion to solid substrata, locomotion, chemokinesis and chemotaxis 27
H.-U. Keller

Peptide and protein chemotactic factors and their recognition by neutrophil leucocytes 53
P. C. Wilkinson

Receptor modulation and its consequences for the response to chemotactic peptides 73
S. H. Zigmond and S. J. Sullivan

Chemotaxis in the cellular slime moulds 89
P. C. Newell

Chemotaxis in the eukaryotic microbes 115
G. W. Gooday

The molecular biology of bacterial chemotaxis 139
G. L. Hazelbauer

Motility and fruiting in the bacterium *Myxococcus xanthus* 155
C. H. Clarke

Index 173

CONTRIBUTORS

Clarke, C., Department of Biological Sciences, University of East Anglia, Norwich, NR4 7TJ, UK.

Dunn, G. A., Strangeways Research Laboratory, Wort's Causeway, Cambridge, CB1 4RN, UK.

Gooday, G. W., University of Aberdeen, Department of Microbiology, Marischal College, Aberdeen, AB9 1AS, UK.

Hazelbauer, G. L., Wallenberglaboratoriet, Uppsala Universitet, Uppsala, Sweden.

Keller, H.-U., Universität Bern, Pathologisches Institut, 3010 Bern, Freiburgstrasse 30, Switzerland.

Newell, P. C., Department of Biochemistry, University of Oxford, South Parks Road, Oxford, OX1 3QU, UK.

Sullivan, S. J., Department of Biology, University of Pennsylvania, Philadelphia, Pa. 19174, U.S.A.

Wilkinson, P. C., Department of Bacteriology and Immunology, University of Glasgow, Western Infirmary, Glasgow, G11 6NT, UK.

Zigmond, S. H., Department of Biology, University of Pennsylvania, Philadelphia, Pa. 19174, USA.

PREFACE

The study of chemotactic reactions has always attracted scientists from many disciplines. Not only does it have an intrinsic fascination as a form of cellular behaviour, but it is now beginning to be possible to approach it at the molecular level. In at least three widely diverse cell types, the enterobacteria, the cellular slime moulds and the mammalian leucocytes, rapid progress is being made in understanding the underlying mechanisms of signal perception and message transduction. Leucocytes have been especially popular because of their importance in disease and have therefore interested not only cell biologists but also immunologists and clinicians.

Chemotaxis is a popular concept to invoke whenever directed cell movements are observed and it is therefore necessary to consider not only the phenomenology, but the *biology* of the response; questions of perception, transduction, and coordinated control. We invited colleagues in a number of fields to discuss the biology of cellular chemotaxis as it relates to their own test organisms. Appropriately the Symposium on which this book is based was part of a meeting organized by the Society for Experimental Biology, appropriately because we are considering, at a cellular level, problems which also face whole organisms.

We hoped that a survey of chemotaxis in a phylogenetically diverse set of cells would illustrate common features of the response and indicate divergences between systems. Perhaps too the effort of communicating with colleagues in adjacent areas of study would crystallize our own thoughts – and serve to stimulate others.

We enjoyed the Symposium and would like to thank our contributors for producing manuscripts which, with little editorial effort, have fitted together without overlap to give the broad coverage of the biology of the chemotactic response that we had set out to achieve.

December 1980

J. M. Lackie
P. C. Wilkinson
Editors for the
Society for Experimental
Biology

JOHN LACKIE & PETER WILKINSON

Introduction

It has been suggested that science progresses by asking *what* happens, *how* things happen, and *why* things happen – in that order. These sets of questions are contained, like sets of Chinese boxes, within each area of science. In the study of cell motility we have reached the stage of asking *why* things happen or, less succinctly but more accurately, what regulates or controls the motile systems. The phenomenon of cell motility and its occurrence in such diverse activities as fertilization, morphogenesis, defensive reactions and neoplastic invasion is well known. In recent years we have learnt much about the mechanism of motility and *how* motility occurs, we have become accustomed to the biochemical and biophysical jargon of actomyosin, tubulin and flagellin, of sliding-filaments and rotating base-plates. One answer to the question of *why* movement occurs seems to be that cells make locomotory responses determined by chemical substances in their environment. The question why cells move in a particular direction may then be answered by the assertion that movement is, in some cases, directed by gradients of diffusible substances, that cells are capable of a chemotactic response. We begin again, therefore, asking what sorts of chemotactic responses occur, how the stimulus is coupled to the response and how the cell regulates its capacity to respond to chemotactic cues.

This collection of papers is an attempt to examine the phenomenon of chemotaxis in the context of cell movement. Given that cells are capable of reacting chemotactically, we can ask how they perceive and respond to the signal and, in passing perhaps, what this response tells us about the control of movement in a more general sense. The most actively studied model systems in this area have been mammalian leucocytes which move for defensive purposes in the inflammatory response, slime moulds which exemplify or model morphogenetic movements, and bacteria in which it is possible to bring the analytical methods of molecular biology into action. Other cells show chemotaxis of course and the capacity of multicellular organisms to respond should not be forgotten, although it will be largely ignored here. Cell biologists have profited much from the studies of animal behaviourists; perhaps the pupils may yet teach.

In examining the phenomenon of chemotaxis at the cellular level we have to put chemotaxis in perspective by considering other possible sources of information in the environment: Graham Dunn (in the first chapter) discusses guidance systems in general and in so doing draws attention to some very fundamental concepts. Such an analytical approach to the information content of the environment is very valuable and this chapter should provoke much interest; it provides some clues as to the ways in which we might analyse the behaviour of cells which are apparently responding in a directional manner. The form of the response to diffusible substances may be complex, involving both kinesis and taxis: Hans-Uli Keller (in the second chapter) discusses these aspects in more detail for the particular case of neutrophil leucocytes. Peter Wilkinson (in the third chapter) continues with leucocytes, looking at the classes of substances which are perceived and the way in which a cell copes with a potentially almost infinite range of specific signals. A first approach to the question of how neutrophils might regulate their chemotactic responsiveness comes from Sally Zigmond and Susan Sullivan (in the fourth chapter): this is an area about which we know very little as yet but which will undoubtedly receive much attention in the near future. Signal transduction and the control of the locomotory machinery in eukaryotic cells are not covered in this volume: we have chosen to define biology as excluding physiology for convenience and because this area is developing so fast that no simple treatment can do it justice. Almost certainly ion movements across membranes are involved and the temptation to make speculative analogies between the chemotactic response, excitation-contraction coupling and stimulus-secretion coupling is almost irresistible and probably very misleading.

Having looked at leucocytes in detail we move to a consideration of the role of chemotaxis in the cellular slime moulds and find conservation of this property in both unicellular and multicellular phases of their life cycle. Peter Newell (in the fifth chapter) reviews this area and introduces the genetic analysis of chemotaxis in slime moulds. Although slime moulds have received a disproportionate amount of attention because of their suitability as a model system the phenomenon is not restricted to them. Graham Gooday (in the sixth chapter) gives a review of the phylogenetic range of chemotaxis in unicellular eukaryotes with particular attention to areas in which preliminary studies have indicated fruitful possibilities for further development. A clue to the possible origin of receptor sites arises from this chapter – the diversion to sensory function of binding sites primarily intended for nutrient transport.

Bacterial chemotaxis is a field often neglected by those interested in eukaryotic systems; although the behavioural response to the gradient differs from that of eukaryotic cells and we might pedantically prefer to consider it klinokinesis rather than chemotaxis, there are important messages for those interested in cell behaviour. Gerry Hazelbauer reviews bacterial chemotaxis (in the seventh chapter) and shows how the elegant and powerful methods of molecular genetics have led to a fairly detailed description of receptor–effector coupling; so much more is known of this in bacterial systems that we ignore them at our peril. Again, receptor sites seem to have been diverted from transport functions in some cases. Lest we become complacent, Colin Clarke (in the eighth chapter) shows us that the Myxobacteria exhibit a complex array of behavioural patterns which are reminiscent of those which are seen in the eukaryotic Myxomycetes (in the fifth chapter). It is sobering to realize how many steps there are in the apparently simple process of chemotaxis; the numbers of gene loci involved are high, even in bacteria, and we should not be surprised if describing the biology of the chemotactic response proves a lengthy task.

To follow our scheme of 'what, how, why' we have had to move from system to system, finding better understanding now in one field, now in another. Only by bringing together these different areas can we hope to produce more coherent descriptions of the response: we hope that this collection of essays will be of some help in stimulating and informing scientists in various disciplines.

GRAHAM A. DUNN

Chemotaxis as a form of directed cell behaviour: some theoretical considerations

It is self-evident that an investigation of the cellular mechanism of any form of directed cell movement would be incomplete without a thorough analysis of the patterns of cell translocation under selected conditions. And yet such analysis is the most neglected aspect of this field of study; there is little consensus of opinion on reliable methods for describing, assaying and interpreting cell behaviour. Perhaps the reason for this is partly that the theoretical basis of such methods is complex and little understood and partly that many feel that the consideration of mathematical models and the classification of responses have little relevance to the real objectives of determining the biological mechanisms.

In this paper I hope to show that some progress can be made towards a consistent theoretical basis while leaving the more formidable mathematical contortions safely in the hands of the mathematicians. Without at least some consideration of behavioural theory such important questions as 'Which properties of the environment could give rise to this form of cell behaviour?' and 'How may we design experiments to distinguish between these two proposed cellular mechanisms?' cannot be answered. As to the relevance of classification, Keller *et al.* (1977) have drawn attention to the prevalent confusion over terminology which is a very real barrier to the communication of ideas and results within the field of chemotaxis. I can offer no easy solutions to this last problem in the form of new and acceptable schemes of classification but I have aimed at least to draw attention to some of the sources of confusion.

Random locomotion

Any study of the patterns of cellular translocation, almost by definition, requires analysis of the deviations from random locomotion. Thus the theory of locomotory behaviour and the design and interpretation of experiments both rest ultimately on the fundamental concept of randomness. Unfortunately, the apparent simplicity of this concept is very deceptive. Furthermore, much of the theory of random movement is still a current problem in mathematics. It is perhaps not surprising, therefore, that the term 'random locomotion' has acquired a notoriety for its misuse and inconsistent use.

Of the more reasonable applications of the term, two have been selected by Keller *et al.* (1977) as being the most useful to students of chemotaxis: '(1) Random direction in relation to the surroundings of the cell i.e. non-directed locomotion . . . (2) Locomotion according to a random walk model . . .'. I shall refer to these as random locomotion type 1 and type 2 respectively. The responses of moving cells to anisotropic properties of their environment are the only source of deviations from random locomotion type 1, and herein lies the usefulness of the term. In contrast, deviations from random locomotion type 2 may also include intrinsic regularities of behaviour which are derived, not from the environment, but from the functioning of the locomotory machinery of the cell.

The exact definition of random locomotion type 1 entails a consideration of the meaning of 'non-directed'. In order to specify a random behaviour in which there is no enhanced displacement of cells along preferential directions within the environment, it is necessary not only that the directions of cell movement, at any one time, are randomly distributed, but in addition their speeds of movement must be unrelated to their directions of movement. For this reason, even with polarized cells whose directions of movement are apparent, the sampling of a population of cells at an instant in time can never yield enough information to distinguish between random locomotion type 1 and directed locomotion. The additional necessary information can only be obtained by sampling the population of cells over a time interval. If, for example, the positions of the cells are recorded at two instants in time, the movement of each cell may be represented by a single displacement vector which has both a direction and a magnitude. Random locomotion type 1 can be defined simply in terms of these displacement vectors as follows: *random locomotion type 1 is that in which the cell displacements during any single time interval are random vectors.* A vector is said to be random if it is drawn in a random direction with a length which is a random variable independent of its direction (Feller, 1966).

Random locomotion type 2, in its most general form, is a proposition that the actual paths taken by the moving cells are generated by entirely random processes. The displacement vectors of a single cell, from consecutive equal time intervals, may be joined end to end in sequence to form a segmented approximation to the actual cell path. The selection of shorter time intervals gives a better approximation. If the displacement vectors are not only random as defined above, but are also independent of each other in sequence, then the linked series of displacements is a pure random walk.

In a pure random walk in two dimensions, the angle of turn, θ, taken after a step is uniformly distributed over the range $-\pi$ to $+\pi$ radians. By integrating $\cos^2(\theta/2)$ it is not difficult to demonstrate that the mean square displacement,

R^2, after two steps each of length a, is equal to $2a^2$. The general result for n steps is more difficult to demonstrate but is just as simple and is given by Feynman, Leighton & Sands (1963) as

$$R^2 = na^2. \tag{1}$$

The probability, P, that a pure random walker will have made a net displacement of length greater than R after n steps of mean square length a^2, is given by Skellam (1951) as

$$P = \exp(-R^2/na^2). \tag{2}$$

Any interdependence between the steps of a random walk is known as internal bias (Patlak, 1953*a*), so called because the bias is derived from the internal properties of the walker and not from its response to the environment (which is external bias). Either form of bias destroys the purity of the random walk. If the direction of a step approximates to that of the previous one (i.e. if the distribution of θ is concentrated around zero) then the internal bias is known as persistence of direction. If the length of a step approximates to that of the previous one then the internal bias is known as persistence of speed. A more general term for either persistence of speed or direction, or both, is persistence of velocity.

A metazoan cell is never a random walker, in the above sense, and it is therefore meaningless to discuss whether a particular cell type is a pure or persistent random walker. The actual path taken by a metazoan cell in culture is a sinuous continuous line which is not made up from discrete segments of straight lines and the term 'random walk', as I have used it so far, applies to various approximations to the path and not to the path itself. It is common sense that if one were to take sufficiently short time intervals, say one second, then the direction or speed of cell locomotion would hardly be expected to change much between one time interval and the next. Therefore, random walks which are better approximations to the true cell path are more likely to show persistence. Indeed, it can be shown that a cell which does not show directional persistence at any level of resolution could never move from the spot. The following informal illustration, based on random walk theory, may help to verify this point.

Consider a cell which, by moving at constant speed, has displaced 100 μm in 100 min. Our minimum estimate of its speed is 1 μm/min. However, if we look at its position every minute and the result is a pure random walk we would expect, from equation (1), each of the 100 steps to be about 10 μm long in order to give a net displacement of 100 μm. Our minimum estimate of its speed becomes 10 μm/min. If we look at its position every 0.01 min and the result is still a pure random walk then each of the 10000 steps will be about 1 μm long and our minimum estimate of its speed becomes 100 μm/min. Since this procedure may be continued indefinitely, the random walks being successively better approximations to the cell path, there must come a stage where a random walk shows persistence of direction unless the cell is moving at infinite speed!

Therefore persistence is a general feature of random walks derived from cell paths in this way, provided that the sampling resolution is high enough, and a definition of random locomotion type 2 should take this into account. For example: *random locomotion type 2 is that behaviour in which, for each cell, the successive displacements during equal time intervals form a random walk with no external bias and no internal bias other than persistence*. Of course, different cell types, or different conditions, may yield different levels of persistence when sampled over the same time intervals.

Expression (2) may be used to predict, from sampled displacements over short time intervals, the probability that a randomly moving cell will have made a net displacement greater than a certain distance after a long time interval. This, of course, assumes that the cell will carry on moving in the same manner. Since the expression is only valid for pure random walks, the sampling intervals must be long enough to ensure that there is no effect of persistence. Gail & Boone (1970) found that persistence is not detectable in a walk derived from 3T3 cell locomotion over five-hour time intervals. From their data, the mean square displacement in five hours is approximately $1.3 \times 10^4\ \mu m^2$. Therefore the probability that a 3T3 cell will make a net displacement of greater than one millimetre in a week (~34 steps) is given by

$$P = \exp\left[-10^6/(34 \times 1.3 \times 10^4)\right].$$

This probability is one in ten which illustrates the important role of directed cell behaviour in achieving relatively large displacements. Neutrophils, of course, can move much faster than 3T3 cells even when they are not in a chemotactic gradient. From the data of Wilkinson & Allan (1978), human blood neutrophils in a uniform concentration of casein show no persistence over three-minute intervals and have a mean square displacement of approximately 2000 μm^2 during each interval. Despite their speed, the probability of such a randomly moving neutrophil showing a displacement of more than one centimetre in a week is less than one in two million!

Before leaving random walks I should point out that the above method of deriving random walks from cell paths is not the only one. The random walk, as a mathematical abstraction, is a sequence of vectors with certain probabilistic properties and the mathematical results do not depend on how the vectors are derived from the real world; they are applicable to any process which produces vector sequences with the specified properties. For a review of some of the varied applications of random walk theory to problems in biological behaviour see

Patlak (1953*a*). In an accompanying paper, Patlak (1953*b*) distinguishes two methods by which the parameters of a walk may be derived from the movement of organisms. In the 'cross-section method', instantaneous observations of position are made at prescribed time intervals as described above. In the 'continuous observation method', determinations are made of how often the organisms turn, how much they turn, and their precise behaviour. In other words, this second method would treat the cell path as if it did consist of straight-line segments linking discrete turns. It appears to have the advantage of allowing a more precise description of behaviour; such things as the cell's speed and frequency of locomotion, magnitude and frequency of turns, and mean free path between turns can all be measured directly from the derived walk, and the type and amount of persistence may be determined. It may even be possible to find a cell path which shows no persistence. However, the method suffers from the serious disadvantage that the observer must subjectively decide where the turns are and the distribution of the derived parameters will reflect not only the cell behaviour but the observer's judgement. Its use may be justified to some extent if there is good reason to believe that the cellular steering mechanism actually works like this, making turns discretely at intervals rather than continuously adjusting the direction of movement. From studies of 3T3 cell locomotion, Albrecht-Buehler (1979*a*) believes that this is the case and that each cell is programmed to move in straight lines in between predetermined turns (although he mistakenly assumes that the detection of persistence in 3T3 cells using the cross-section method (Gail & Boone, 1970) supports this contention). Even if this mechanism applies, the fact remains that the resulting cell path is not precisely defined into straight-line segments and the practical difficulty of detecting low-angle turns inevitably imposes a serious bias on the estimated angular distribution of turns and the distribution of the interval between turns. In the following sections I shall therefore consider only the objective cross-section methods of sampling cell behaviour.

Non-random locomotion

Patterns or regularities of cell behaviour are detected by testing hypotheses of non-randomness. Unfortunately, the theory of randomness does not help much in formulating hypotheses of non-randomness. In fact, the randomness of any real process can only be described in terms of the hypotheses of non-randomness which have been tried and rejected. Infinitely many hypotheses of non-randomness are possible and the rest of this paper, in common with almost any other discussion of cell behaviour, is concerned with which criteria may be used for selecting informative hypotheses of non-randomness. Among these, of course, are the criteria based on direct observation of cell behaviour. But useful criteria may also be provided by the analytical approach: an attempt to resolve behavioural patterns into simpler components which may reflect the underlying mechanisms, or by a consideration of the information content of environments which elicit regularities of behaviour.

Internal bias

In a uniform isotropic environment (that is one without directional information) where the cells are not allowed to interact with each other, they will show random locomotion type 1 and the only possible departures from random locomotion type 2 will be those of internal bias. Apart from persistence of velocity, other regularities of speed or direction or both also constitute internal bias. For example, the changes in locomotory parameters which accompany the mitotic cycle in dividing cells or which accompany any progressive deterioration of the cells in culture will give rise to internal bias. Any regular pattern of turning or stopping and starting also constitutes internal bias. Since this paper is concerned with directional behaviour, internal bias merits only a brief mention, but one hypothesis of internal bias is of interest here. Albrecht-Buehler (1977) has proposed that the sequence of displacement vectors of a 3T3 cell is not independent of the displacement sequence of its sister cell. He has called this hypothesis ‘predetermined guidance’ and suggested that such predetermination may play a role in morphogenesis, but the postulated relationship between the behaviour of sister cells, which is the essence of this interesting hypothesis, has yet to be tested rigorously.

Simple kinesis

On mechanistic criteria, the simplest theoretical locomotory response of a cell to its environment is one in which the magnitude of cell motility is related to the magnitude of some property of the environment. This is a simple kinesis. Later I will consider more complicated forms of kinesis and their relationship to mechanisms in which the direction of cell locomotion is related to both the magnitude and direction of properties of the environment; among such mechanisms is the taxis. Since the distinctions of interest between different types of behaviour eventually lie at the level of the mechanisms, I will tend in this paper to prefer the theoretical mechanistic distinctions as the primary definitions of responses and to deduce the resulting behaviour. If distinctions which are consistent with the mechanistic ones could be made at the behavioural level then the analysis of a behaviour pattern could give some information on the nature of the cellular receptors and this, of course, is the main aim of behavioural studies. Such behavioural distinctions are unfortunately not always easy to make, especially using just one type of experiment, and this difficulty is compounded if the distinction is attempted purely on the grounds of the final distribution of cells, as in assays of chemotaxis using the Boyden chamber.

In classical analysis, definitions are usually based on behavioural distinctions, and mechanistic distinctions are only resorted to in case of difficulty. In a classification of the behaviour of free organisms (Ewer & Bursell, 1950), the term orthokinesis is applied to behavioural responses where the speed or activity (frequency of locomotion) is dependent on some property of the environment and klinokinesis is applied to those where the rate of change in direction (frequency and/or magnitude of turns) is affected. A simple kinesis may be either a pure orthokinesis or a pure klinokinesis, or a mixed kinesis. When applied to cells, these terms are unwieldy as a behavioural distinction for several reasons. Firstly, the precise meaning of the terms, especially klinokinesis, is not commonly agreed and several interpretations are present in the literature. Secondly, simple estimates of speed or rate of change in direction, obtained using the cross-section method, are biased, that is the mean value obtained for either parameter depends upon the duration of the time intervals used for the measurements. Thus the description of cell behaviour in terms of orthokinesis or klinokinesis can only apply if the sampling intervals are specified. Thirdly, the alternative continuous observation method, which could yield unbiased estimates in ideal circumstances, is insufficiently objective when applied to the wiggly paths of cells, as discussed earlier. Fortunately, for some purposes it is not necessary to make this behavioural distinction. Simple kinesis behaviour may be analysed in terms of cell motility which is a combination of the above parameters.

Gail & Boone (1970) have proposed an unbiased measure of cell motility, the augmented diffusion constant, which may be obtained using the objective cross-section method of analysis. In a persistent random walk the estimated mean square displacement, $\langle R^2 \rangle$, during a long time interval, t, is given by $\langle R^2 \rangle = 4D^* t$ where D^* is the augmented diffusion constant. If the experiment is restricted to the use of small values of t, then D^* may still be estimated using a more refined equation provided by Gail & Boone.

If an environment is non-uniform with respect to a property which elicits a simple kinesis response, then the motility of a cell will depend on its position within the environment. If the environment is also a closed culture chamber where the limits of the chamber 'reflect' the cells as opposed to stopping them, the distribution of cells will eventually reach an equilibrium where the local density of cells is inversely related to their local motility.

Imagine that the chamber is divided into equal small square elements. With purely kinetic behaviour the emigration of cells from a small element during a finite time interval is, on average, the same in all directions. If the number of cells which leave each element during the interval is the same for all elements, this is obviously balanced by the number which enter each element and the system is in a steady state. On average there is no net flow of cells across any boundary within the system and the cell density at any point does not change with time. Now, in elementary diffusion theory, Fick's law states that the number of particles leaving a small element during a finite time interval is proportional to the concentration of particles within the element and to the diffusion constant, D, within the element. For the persistent behaviour of cells, D may be replaced by D^* provided that the dimensions of each element are large enough to accommodate cell paths which may be approximated by pure random walks. Since the number of cells leaving any element is constant for all elements in the steady state, the concentration of cells within any element is inversely proportional to the local D^* in the steady state. This condition is a stable dynamic equilibrium since any disturbance of the cell distribution will result in a net flow of cells which will tend to restore the steady state. This result should only be regarded as approximate because of the two conflicting assumptions: that each element is small enough that D* may be regarded as constant within the element, and that each element is large enough to accommodate a pure random walk of several steps.

With a simple kinesis, once the equilibrium condition has been reached there is no net directional flow of cells and, although the cells may show accumulation, that is their distribution may not be uniform, they will nevertheless conform to random locomotion type 1. The difficulties arise with distinguishing between a kinesis which has not reached equilibrium and a taxis. Both behaviour patterns show net flows of cells which, if the cell density was initially uniform, lead to accumulation, and both show deviations from random locomotion type 1. The question arises: from where, in the case of a kinesis, does the directional behaviour come? Consider a closed chamber with reflecting walls where the cell motility is constant throughout the chamber. If the cells are initially aggregated in one side of the chamber, they will tend towards a random uniform distribution as equilibrium is approached and thus there will be a net directional flow of cells. However, as the cells diffuse from the centre of aggregation, there will be no net flow of cells until the nearest wall of the chamber is encountered. It is the reflection of cells from this wall which imposes the directionality on the otherwise random diffusion of cells. Similarly, with a kinesis where the motility is greater on one side of the chamber and the cells initially have a random uniform distribution, the walls in the region of higher motility will reflect more cells and thus impose a directionality on the behaviour.

In order to show the relationship between motility and the classical parameters of orthokinesis and klinokinesis I shall use an expression which relates the diffusion constant, D, to a pure random walk. This is given by Gail & Boone (1970) as

$$R^2 = 4Dt \qquad (3)$$

where R^2 is the mean square displacement and t is the time taken by the pure random walk. Note the similarity between this expression and the one relating

D^* to a persistent random walk. In fact, in the limit where persistence is zero, D^* is equal to D (Gail & Boone, 1970). The right-hand sides of expressions (1) and (3) may be equated to give

$$4D = na^2/t. \quad (4)$$

In the case where the step length, a, of a pure random walk is constant, the actual distance travelled by the walker along its path is na and its speed is therefore na/t. The frequency of making turns with distance travelled is n/na or $1/a$. Therefore expression (4) may be interpreted as *the diffusion constant of a pure random walker is proportional to the speed of the walker divided by the frequency of making turns with distance travelled.* This result only applies if the frequency of making turns is referred to distance and not to time.

In a pure orthokinesis, the speed is determined by the environment and the frequency of turning with distance travelled is constant; in other words, the spatial characteristics of the path are constant. Thus the diffusion constant of an orthokinesis is directly proportional to speed. Similarly, with a pure klinokinesis the speed, or temporal characteristic of the path, is constant and the diffusion constant is inversely proportional to the frequency of turning with distance travelled. For a population of pure random walkers in a closed system, the local density of walkers at equilibrium is proportional to the local frequency of turning with distance divided by the local speed. This is illustrated using a simulation of pure random walks in Fig. 1.

Thus the classical parameters used for describing kinesis in organisms may easily be used to describe the characteristics of pure random walkers but they are not so easily applied to metazoan cell behaviour. If the cross-section method is

Fig. 1. Simulation of a simple mixed kinesis. 1000 'cells' are initially distributed at random in two dimensions in a rectangular chamber with reflecting walls. The diagrams, from left to right, represent the distribution of cells along the length of the chamber after 1000, 2000 and 3000 time intervals respectively. During each time interval, each cell in the left half of the chamber makes one step of a pure random walk whose length is 0.01 of the length of the chamber; each cell in the right half makes one step of length 0.02. Thus the cells in the right half have twice the speed and half the frequency of turning with distance of those in the left half and consequently they have four times the diffusion constant of the left-hand cells (see text). The arrows indicate the predicted population of cells in either half when equilibrium is reached.

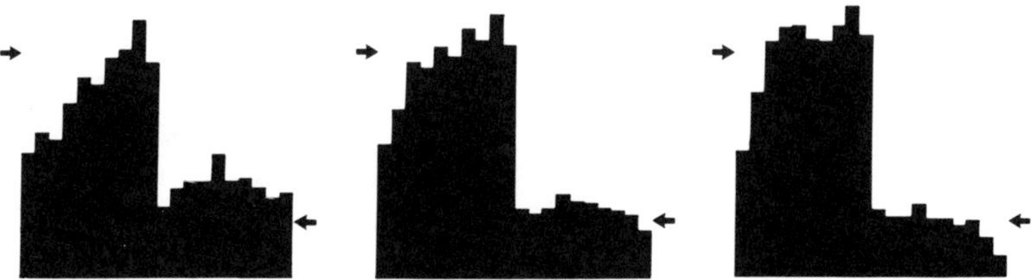

used for analysing cell behaviour, the derived walk may be described in terms of speed and rate of change in direction with distance, but the values obtained may change when the sampling interval is changed. There is the further complication that the rate of change in direction with distance is a function not only of the frequency of turns of the derived walk but also of the persistence in direction of the walk. The only possibility of an absolute distinction between orthokinetic and klinokinetic behaviour of cells is to use the continuous observation method, with all its attendant difficulties. Despite the subjective nature of this method, it should be possible to distinguish a fairly pure orthokinesis, where the spatial characteristics of the path are independent of the environment, from a fairly pure klinokinesis, where the temporal characteristics of the path are independent of the environment. Whether either of these extreme forms of simple kinesis occurs in the behaviour of metazoan cells has yet to be established.

To summarize the practical aspects of this discussion of simple kinesis in metazoan cells: a simple kinesis may be identified and characterized by examining the cell behaviour in environments which are uniform with respect to the kinetic factor. Motility curves may be plotted using the method of Gail & Boone (1970) at different levels of the factor. In each case, the cells should be given time to adapt to the environment before assaying the motility. If there is no evidence of adaptation, then the motility data may be used to predict the equilibrium distribution of cells in a stable environment which is non-uniform with respect to the kinetic factor. It is usually impractical to expect cells in short-term culture to achieve equilibrium but, if the distribution at any stage tends to diverge significantly from the predicted equilibrium condition, then it is probable that the response also has a tactic component which is induced by gradients in the distribution of the active factor.

Taxis

There is the possibility, therefore, that a simple chemokinesis component and a chemotaxis component may coexist in the response of cells to chemical gradients and this seems to be realized in the response of leucocytes to chemotactic factors (Wilkinson & Lackie, 1979). A similar possibility exists with other forms of taxis such as the haptotaxis (Carter, 1965) of L cells on gradients of substratum adhesiveness. Although Carter did not analyse the kinesis component of the response, Gail & Boone (1972) have demonstrated a simple kinesis response to substratum adhesiveness in 3T3 cells. With some forms of taxis, however, a consideration of the information content of the environment rules out the possibility of a simple kinesis component. For example, certain non-metazoan cells have been found which exhibit a taxis response in a uniform magnetic field (Blakemore, Frankel & Kalmijn, 1980), in a uniform gravitational field (Roberts, 1970) or in a uniform field of luminous flux (Holmes, 1903). As with a chemical concentration gradient these fields provide directional information to the cell but, unlike a concentration gradient, no information is available to the cell on its position within the environment and it is therefore inconceivable that the response has a simple kinesis component. The difference is that the magnetic, gravitational and luminous flux fields are vector fields whereas fields of chemical concentration, substratum adhesiveness, temperature and the intensity of illumination at a surface are scalar fields. In mathematical terms the distinction is not very fundamental since the derivative of a scalar field is a vector field and, conversely, just as an electric field may be considered a gradient of electric potential, so many vector fields may be thought of as gradients of potentials. The distinction of biological importance is whether the potential can elicit a kinesis response, and physics helps to the extent of indicating that in some cases the potential is an abstract concept quite incapable of having a biological effect. Even if the potential is a real scalar property, it need not have a biological effect, and in a hypothetical pure chemotaxis, it is the vector field, the magnitude and direction of the concentration, to which the cells respond. Similarly, in all cases, a tactic environment may be described as a vector field.

Temporal mechanism of taxis

A kinesis response is often not simple, it may show adaptation. In a uniform environment where a scalar property, such as the concentration of an active substance, suddenly changes with time from one magnitude to a new constant magnitude the cells may respond not by changing to a new constant level of motility, as with a simple kinesis, but by showing a transient change in motility. If the motility reverts to its initial level after the transient wave then the kinesis shows perfect adaptation. This behaviour implies the existence not only of a cell receptor, as with a simple kinesis, but also of a cell 'memory' (Gunn, 1975). The memory, of course, need only be very simple, possibly the concentration of some substance within the cell. An adapting cell continuously compares the concentration of the active substance in the environment with its memory of the concentration a short time ago; if the concentration has not changed over the short time interval then the motility remains at or reverts to the resting level but if there is a relative change in the concentration the level of motility either increases or decreases appropriately. With rapid perfect adaptation, the cell motility is only maintained at a new constant level if the concentration is either increasing or decreasing continuously.

The effect of rapid and perfect adaptation is that the cell responds not to the absolute concentration but to the rate of change in concentration with time. If the environment is a gradient of concentration which does not change with time, the response takes effect nevertheless since the cell, by moving through the environment, experiences changes in local concentration with time. The exact nature of the effect may be best understood by considering the classical components of this adapting kinesis: orthokinesis and klinokinesis. If the mechanism is a pure klinokinesis with adaptation, that is if the cell speed does not change as a result of the kinesis then, ignoring random changes in speed, the rate of change in concentration with respect to time is equivalent to the rate of change in concentration with respect to the distance travelled by the cell. In other words the cell responds to the spatial concentration gradient and this is the exact requirement of a taxis. By adapting perfectly the cell has ceased to be able to respond to the scalar magnitude of the concentration and is only able to respond to the vector quantity of the gradient.

A more specific example may help to clarify this point. Consider a mechanism of klinokinesis with perfect adaptation such that a sudden increase in local concentration of active substance leads to a transient decrease in the rate of change in direction of the cell with respect to distance travelled; a cell which happens to be moving up a concentration gradient will tend to carry on moving in that direction since the local concentration will be increasing continuously, whereas a cell which happens to be moving down gradient will tend to turn away from that direction. Fig. 2 shows that an adapting klinokinesis can be a very efficient form of directional behaviour. Only the klinokinetic component of an adapting kinesis can contribute to this taxis behaviour since a pure orthokinesis, whether with or without adaptation, produces a cell path with spatial characteristics which are independent of the environment; it is only the speed with which the cell traces out the spatial path which is affected.

It has been appreciated for some time that a klinokinesis with adaptation can give rise to a continuous directed movement, very different from the equilibrium distribution which results from a simple klinokinesis (Fraenkel & Gunn, 1940; Ewer & Bursell, 1950). In fact the original meaning of klinokinesis, as used by Gunn, implies an adaptation (Gunn, 1975) and the reaction was thought to account for the directional response of *Dendrocoelum lacteum* to gradients of light intensity (Ullyot, 1936). As Gunn (1975) points out, this early interpretation of the behaviour of these flatworms is now suspect but more recent work on the response of bacteria to chemical gradients (see for example Macnab & Koshland, 1972, and Hazelbauer, this volume, pp. 139–141) is thought to conform closely to a klinokinesis with adaptation. In modern usage this behaviour is more often referred to as the temporal mechanism of taxis.

Variations in the parameters of kinesis with adaptation can lead to an almost endless variety of patterns of cell behaviour. In the case of klinokinesis behaviour in a gradient, if there is no adaptation the cells will accumulate preferentially at the end of the gradient where the motility is lowest (i.e. where rate of change in direction is highest) until an equilibrium distribution is reached. If there is perfect and rapid adaptation the cells will flow steadily and continuously to the opposite end of the gradient! This partially accounts for the frequent confusion in the use of the terms positive and negative kinesis. If the adaptation is imperfect or if it is not very rapid then various intermediate patterns of behaviour may result; the mathematics of predicting the patterns now becomes quite formidable. Finally, if the adaptation is perfect and instantaneous then the situation is the simplest of all: the cells show no response to the gradient.

Spatial mechanism of taxis

In contrast to the temporal mechanism of taxis it is possible to envisage a mechanism whereby a cell could instantly obtain some directional information

Fig. 2. Simulation of an adapting klinokinesis. 100 'cells' are initially distributed at random in two dimensions in a rectangular chamber with reflecting walls. The diagrams, from left to right and top to bottom, represent the distribution of cells along the length of the chamber after 0, 50, 100 and 150 time intervals respectively. The environment is a stable linear gradient which takes the value 0 on the extreme left of the chamber and 1 on the extreme right of the chamber. During each time interval, each cell makes one step of a persistent random walk whose length is 0.01 of the length of the chamber. The turn, at the end of each interval, has a random normal distribution with zero mean and a standard deviation which is determined by the change in the cell's local environment during the interval. In the extreme case, where the change in local environment is +0.01, the standard deviation of the turn is zero and, for the opposite extreme where the environmental change is −0.01, the standard deviation of the turn is 1 radian. Intermediate changes in environment produce intermediate deviations.

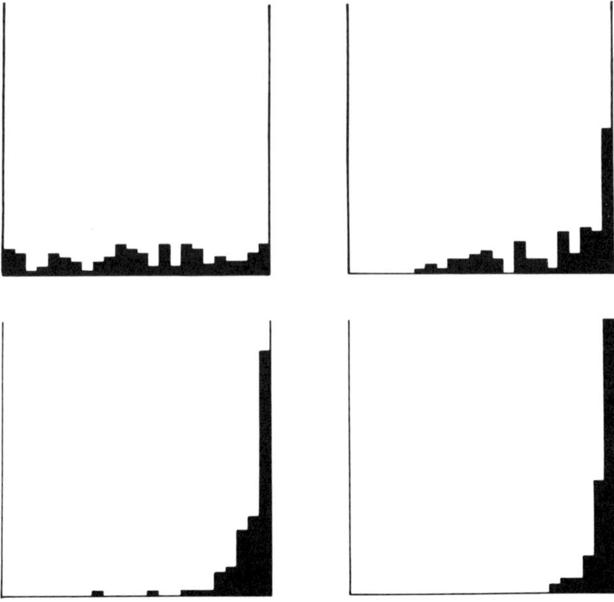

from a vector field without having to move through it. A memory is not theoretically required by such a mechanism but, if a single receptor is incapable of responding to the direction of the vector field, then two or more spatially separated receptors are required. Zigmond (1974) has demonstrated that such a spatial mechanism probably applies to the chemotaxis of leucocytes. She observed that, before translocation commences, the initial orientation of leucocytes in a gradient of chemotactic factor is significantly biased in an up-gradient direction; a temporal mechanism would be expected to give a random distribution of initial orientations.

Wilkinson & Lackie (1979) point out that nearly all chemotactic factors for leucocytes can elicit a simple kinesis response with little or no adaptation. It is not known whether this is an inevitable side effect of the taxis mechanism or whether there are two separate mechanisms, one for the simple kinesis and one for the taxis, which, for biological reasons, are both sensitive to the same range of substances. The observations of Keller, Wissler, Hess & Cottier (1978), that a particular purified chemotactic factor does not elicit kinesis, suggests that two distinct mechanisms are more probable. Zigmond & Sullivan (1979) found another effect of certain chemotactic peptides on leucocyte locomotion which may well turn out to be an invariable property of leucocyte chemotactic factors. They observed that a sudden increase in the concentration leads to a transient increase in the spreading and ruffling activity of the leucocytes whereas a sudden decrease in concentration leads to a transient withdrawal of lamellar protrusions and a transient blebbing activity. As with a kinesis with adaptation, it is possible that such a transient response could be extended over a much longer time course if the concentration were rising or falling continuously. However, Zigmond & Sullivan found that the movement of leucocytes up a chemotactic gradient was never sufficiently rapid to maintain the response. They proposed a mechanism by which adapting ruffling activity, despite its transient nature, might yet be an essential part of the chemotactic response but this, in common with many other models of spatial chemotaxis, would probably require some complicated means of amplification in order to explain the observed response to very small differences in concentration across the cell.

An alternative simpler explanation has occurred to me which may at least be worth investigating (see also Zigmond & Sullivan, this volume). It is based on the idea that the spatial mechanism of leucocyte chemotaxis may not be very different from a temporal mechanism. In a temporal mechanism the receptor is assumed to be fixed in relation to the cell and its movement is entirely due to the translocation of the cell but an adapting receptor which moves in relation to the cell could lead to a behaviour which is similar to that due to a spatial mechanism. Assume that an adapting receptor is situated at the active edge of ruffles in leucocytes and that its response is to control the local rate of protrusion. Provided that the concentration at the receptor is increasing rapidly and continuously, protrusion is stimulated but, if the concentration is decreasing rapidly and continuously, the result is withdrawal. A step-like increase in uniform concentration would lead to a transient ruffling activity of the whole cell surface, as observed by Zigmond & Sullivan (1979). In a concentration gradient, the protrusion of a ruffle or lamellipodium up-gradient might be self-reinforcing, at least until it is limited by other factors such as the local availability of cellular material for protrusion. Whether this protrusion is reinforced depends on whether the change in concentration at the receptor is sufficiently rapid; and this would in turn depend not on the speed of locomotion of the cell as a whole but on the speed of protrusion of the ruffle which might be several times greater. Similarly, the protrusion of a ruffle or lamellipodium down-gradient might be self-inhibiting. The mechanism would operate by restricting protrusion to an up-gradient direction and thus determining the polarity of locomotion of the cell. It may obviate the necessity of postulating messengers within the cell since there is some evidence that the edge of a lamellipodium is also the site of assembly of material for protrusion (Dunn, 1980); all that is needed is that this assembly is directly sensitive to the rate of change in concentration of chemotactic factor.

I have included this model, not as a serious proposition for a complete mechanism of chemotaxis in leucocytes, but because the possibility of moving receptors appears to have been largely ignored as a partial explanation of taxis in metazoan cells. Also it demonstrates a possible close relationship between the temporal and spatial mechanisms of chemotaxis.

Guidance

A taxis is not the only form of directed cell behaviour which has been recognized. In the earliest days of tissue culture Harrison (1912) concluded that 'the behaviour of the cells with reference to the surface of the coverslip and spider web shows not only that the surface of a solid is a necessary condition (for cell movement) but also that when the latter has a specific linear arrangement, as in the spider web, it has an action in influencing the direction of the movement as well as upon the form and arrangement of the cells'. Since then Paul Weiss (see Weiss, 1959) has gathered together a number of instances of this form of directional behaviour and has termed them 'contact guidance'. The essential difference between a taxis and a contact guidance does not lie in whether it is the fluid phase or the solid phase of the environment which imparts the information to the cell but in the different symmetries of the two patterns of behaviour. For this reason I will drop the adjective and refer to this group of responses as guidances. In the case of two-dimensional environments, in which cell locomotion has only two degrees of freedom, a taxis is a unidirectional response in which movement in one direction predominates over all others whereas a guidance is a bidirectional response in which equal movement in two opposite directions predominates over all others. Thus, on biological criteria, there are at least two groups of hypotheses of directed locomotion which will need different testing routines. In testing for a taxis we may look for a continuous net displacement of the cell population but a guidance will show no net displacement unless other factors are involved. In order to decide if these are the only two useful groups of hypotheses about directed behaviour we must turn to theoretical criteria and one such is the possible information content of the environment.

A simple kinesis is a response solely to the magnitude of a property of the environment which is a scalar quantity such as chemical concentration in the fluid phase or adhesiveness of the substratum. A taxis is a response to the magnitude and direction of a property of the environment which is a vector quantity and, as we have seen, linear gradients of concentration or adhesiveness are uniform vector fields where the magnitude is the steepness of the gradient and the direction, 'up-gradient', is uniquely defined. A guidance is a response to a property such as curvature of the surface of the substratum (Curtis & Varde, 1964; Dunn & Heath, 1976) and cell guidance is commonly demonstrated on a cylindrical surface such as a fine glass fibre. Here the quantity of surface curvature, ignoring the ends, is the same everywhere on the surface so that the surface is a field of uniform curvature. But it is not a uniform vector field because no single direction is uniquely defined. Starting from a point on the surface, there are two opposite directions which lie precisely along the cylinder and, at right angles to these, two opposite directions lie precisely around the cylinder. Surface curvature is in fact a physical quantity of yet a higher order of complexity and is sometimes simply called a symmetric tensor. A two-dimensional symmetric tensor such as surface curvature may be expressed as an axis of orientation and two magnitudes. Thus, to describe the surface curvature at a point on a cylinder it is sufficient to define the axis of measurement parallel to the main axis of the cylinder and to state the magnitude of curvature along the cylinder (which happens to be zero in this special case) and the magnitude of curvature, at right angles, around the cylinder. It is a feature of such symmetric tensors that the curvature at any single point on a surface, no matter how complex, may be described simply as an axis and two magnitudes.

The reported cases of guidance behaviour have the common feature that some predominant property of the environment may be described by a symmetric tensor. Early experiments of Weiss (1929) demonstrated cell guidance in stressed plasma clots and here mechanical stress is a tensor. Elsdale & Bard (1972) found that hydrated collagen lattices, aligned by drainage, would elicit cell guidance and fibrillar orientation is a tensor, as is the orientation of fine scratches in the lightly abraded surfaces used by Ohara & Buck (1979) which also cause guidance. This does not imply that the mechanism of guidance is different in each case; it has been suggested, for example, that the response to fibrillar orientation may use a similar cellular mechanism to the response to surface curvature (Dunn & Ebendal, 1978). Also several tensor properties of a single environment may be interdependent; thus stress leads to strain which may lead to orientation in a fibrillar matrix which may result in an anisotropic elasticity. There are intriguing opportunities for experiment and there may be separate responses, involving different mechanisms, to unresearched tensor properties of the environment such as anisotropic elasticity.

Guidance behaviour is more markedly different from taxis behaviour in three-dimensional environments such as in collagen gels. A taxis is much the same in three dimensions as it is in two; one direction is still uniquely defined by a three-dimensional uniform vector field, which may still be described by a magnitude and a direction, the direction requiring two numbers to be expressed in three dimensions. The resulting cell behaviour is still such that movement in one direction predominates over all others. However, a uniform three-dimensional symmetric tensor field requires a system of three axes (which may be expressed by three numbers) and three magnitudes to describe it and, correspondingly, the cell motility may differ in each of the three dimensions. The cell motility itself may be described by an anisotropic diffusion constant which is also a symmetric tensor quantity. Thus in a stretched collagen gel, bidirectional cell movement may predominate along one axis, the axis of stretch, and be suppressed in each of the two axes at right angles to the stretch. This is also the general pattern of behaviour seen in collagen gels which are aligned by drainage (Elsdale & Bard, 1972). On the other hand, in a compressed collagen gel, movement along the axis of compression may be subdominant to movement in the other two axes. Although this possibility has not yet been explored, it seems likely, in my experience, that the behaviour of fibroblasts in the early cornea of chick embryos approximates to it. It may be useful to differentiate between these two patterns of three-dimensional guidance as 'prolate guidance' and 'oblate guidance' respectively by analogy with the terminology for ellipsoids.

Assaying taxis and guidance

The most direct approach to testing a population of cells for deviations from random locomotion type 1 in two dimensions is to plot a vector scatter diagram of the cell displacements over a reasonably short time interval. On such a diagram plotted using polar coordinates, the displacement of each cell is represented by a point whose distance from the origin is the magnitude of the displacement and whose direction from the origin is the direction of the displacement (Fig. 3(*a*)). In practice this may not be a very convenient or sensitive approach to assessing taxis or guidance but it does emphasize the underlying principle of most methods.

If a pure taxis is suspected, the simplest hypothesis of non-randomness is that there is a significant mean cell displacement. In some cases, a simple kinesis might give a small but significant mean cell displacement but, if a kinesis is present, it is often possible to design the experiment to avoid its effects. By taking arbitrary orthogonal axes centred on the origin of the scatter diagram (Fig. 3(*a*)), the displacement due to each cell may be given the coordinates (x,y) and the mean displacement has the coordinates $(\bar{x}, \bar{y})$. An approach to testing for the significance of this mean is to consider all the individual displacements, joined end to end, to form a walk. The order of steps in a walk is irrelevant to the net displacement and so it is possible to formulate the null hypothesis: that the net displacement of the walk is no greater than would be expected if it were the net displacement of a pure random walk in two dimensions. Skellam's formula, which I used earlier in this paper, gives the probability that a pure random

Fig. 3. Method of testing for a taxis or guidance. (*a*) The black circles represent the displacements in this scatter diagram of 20 hypothetical cells (produced using normally distributed pseudo-random number pairs to simulate a guidance). The origin of the displacements is indicated by a large circle and the position of the vector mean by +. In this case the mean displacement is not significantly different from zero (see text) but a taxis would be expected to show a significant mean displacement. (*b*) The first stage in transforming the scatter diagram to test for a guidance is to reverse the sign of the displacements which lie in the lower half of the diagram. The displacements which have been reversed are now represented by open circles. (*c*) The angle between each displacement and the vertical axis is now doubled to complete the transformation. The new vector mean, indicated by +, is now very significant as would be expected with a guidance.

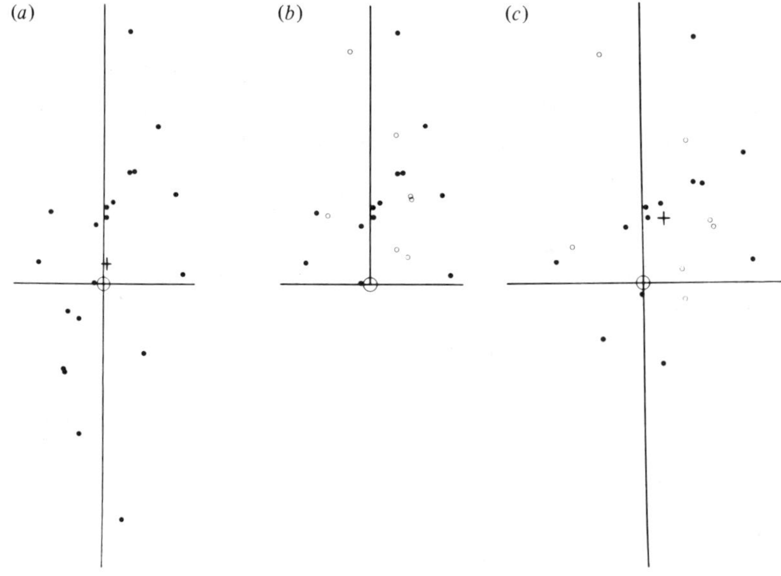

walk of n steps, of mean square length a^2, will have a total displacement of greater than R

$$P = exp\ (-R^2/na^2). \tag{2}$$

For the cell displacements

$$R^2 = (\Sigma x)^2 + (\Sigma y)^2 \text{ and}$$
$$na^2 = \Sigma x^2 + \Sigma y^2.$$

Using the artificial data of Fig. 3(*a*), $P = 0.49$; therefore there is no significant net displacement.

With a pure guidance response, there is no significant mean displacement and the artificial data of Fig. 3 were intended to simulate a guidance response. The simplest hypothesis of non-randomness is that the component displacements in both directions along some particular axis are significantly different from those along the orthogonal axis. An assay system should therefore ignore the sign of each displacement, that is it should ignore whether a cell has travelled from A to B or from B to A. This may be done by employing a technique developed for the analysis of axial data (Mardia, 1972). By ignoring the sign of each displacement, it may be given a direction in the range $-90°$ to $90°$ with respect to an arbitrary axis, such as the y-axis of Fig. 3(*b*). The open circles now represent the displacements which were below the x-axis in Fig. 3(*a*). If the angles are doubled, as in Fig. 3(*c*), the result is a vector scatter diagram which again occupies the full 360° but now opposite directions in this transformed diagram represent directions which were at right angles in the original scatter diagram of Fig. 3(*a*). A significant mean displacement in the transformed diagram means that there was a significant bidirectional displacement in the original diagram. The angle of the transformed vector mean must be halved to give the original axis of predominant bidirectional displacement. By applying Skellam's formula to the data of Fig. 3(*c*), $P = 0.0013$, which indicates a significant guidance.

Scatter diagrams may also be useful in testing whether there is a significant taxis or guidance along a known field axis. Since the components of the cell displacements, in the direction of the field axis, may each be described by a single positive or negative magnitude, standard statistical tests may be employed to determine the significance of the mean. A taxis may be described simply in terms of the mean and standard deviation of the component displacement, in the field direction, during the time interval, and a guidance could be described by applying the same method to the transformed scatter diagram.

Boundary responses

So far the discussion of taxes and guidances has been confined explicitly or, in some cases, implicitly to environments where the properties change gradually from place to place. However, the response of organisms and cells to a sharp boundary in some environmental property has received much attention in the literature and Jennings (1906) introduced such terms as 'shock reaction' and 'avoiding reaction' to express the boundary responses of lower organisms. The popularity of boundary experiments no doubt is due, in part, to their usefulness as an investigative tool and also to the frequent occurrence of boundaries in natural environments. The range of possible boundary conditions is enormous and space permits only a brief discussion here of the simpler ones: boundaries between two isotropic environments.

It has long been recognized that boundary responses may be due to the same mechanisms which produce the responses to gradients and anisotropic fields. Avoiding reactions in different organisms have been attributed variously to a special case of klinokinesis with adaptation (Ullyot, 1936) or to a special case of spatial taxis (Bursell & Ewer, 1950). Further research on metazoan cell behaviour will probably show that the three basic response mechanisms, a simple kinesis, a taxis and a guidance, can each give rise to distinctive behaviour patterns at appropriate boundaries. Simple models for the three mechanisms allow only a limited prediction of the behaviour patterns at boundaries and the details of the behaviour patterns, found by experiment, should therefore prove extremely useful for refining our knowledge of the mechanisms.

I have already discussed the behaviour due to a simple pure kinesis, in which the cell motility on one side of a boundary differs from that on the other side. The proposed relationship at equilibrium, between the augmented diffusion constant and the cell density, rests on the assumption that crossing the boundary has no effect on the cell other than a change in the level of motility. With some possible mechanisms of kinesis behaviour this assumption is not justified. For example, Albrecht-Buehler (1979*b*) has suggested that if the cell is treated as an extended object, which changes its speed on crossing the boundary, it should also change its direction in accordance with Snell's law of refraction.

With a pure taxis, whether temporal (klinokinesis with perfect adaptation) or spatial, in which there is no response to the scalar magnitude of the environmental property, the resultant cell behaviour should be the same on either side of the boundary. The boundary itself may be likened to a narrow vector field and will activate the taxis mechanism; the effect on a cell of meeting or crossing the boundary will depend on the side from which the cell approaches it. The boundary may thus accelerate a cell which crosses it in one direction and retard or stop a cell which approaches it from the other. Depending on the exact mechanism of the taxis, there may be a change in direction of the cell on traversing the boundary, or the cell may stop at the boundary and eventually move off in a new direction, or it may tend to follow the boundary.

Tensor fields may be isotropic, for example, a plane surface or a spherical surface is isotropic with respect to surface curvature. A boundary between two similar isotropic fields may be likened to a narrow anisotropic tensor field and, if the cell has a guidance mechanism for responding to the tensor property, this will be activated by the boundary. The cell behaviour will be the same on either side of the boundary but, in this case, the response will be independent of the side from which the boundary is approached. As with a taxis boundary the exact nature of the response to a guidance boundary will depend on the exact nature of the guidance mechanism.

The study of the behaviour of metazoan cells at boundaries has not been as extensive as in the lower organisms. However, it does seem at least that there are fairly clear-cut examples of a boundary taxis combined with a simple kinesis and of a boundary guidance. Carter (1965) has studied the behaviour of L cells on thin films of cellulose acetate which are rendered more adhesive, to varying extents, by the controlled deposition of palladium. As I mentioned earlier, on steep gradients of adhesiveness he found that the cells move up the gradient in a markedly directional manner and he termed this behaviour 'haptotaxis'. The parallel observations of Harris (1973) using 3T3 cells on similar gradients were in general accordance with Carter's, although Harris found much less directionality in the locomotion. This raises the possibility of a kinesis component and Gail & Boone (1972) have demonstrated a simple kinesis response to substratum adhesiveness in 3T3 cells.

At adhesive boundaries both Harris and Carter are agreed that the cells move across the boundary towards the substratum of higher adhesiveness with apparent facility but rarely, if ever, move across in the opposite direction unless forced to do so by population pressure. Since there is considerable cell motility on the areas of higher adhesiveness, as well as on those of lower adhesiveness, this failure of the cells to cross the boundary in one direction suggests that the accumulation of cells on the areas of higher adhesiveness is not an extreme case of kinesis equilibrium but is due to a boundary taxis. Furthermore, Albrecht-Buehler (1979*b*) has observed that 3T3 cells often tend to follow an adhesive boundary, which supports the supposition that the response is more elaborate than a simple kinesis.

A special case of the boundary response to substratum adhesiveness is where two parallel boundaries enclose a narrow strip of higher adhesiveness. Cells which become confined to the narrow strip are restricted to bidirectional locomotion and examples of this behaviour have been referred to as contact guidance (Weiss, 1959; Carter, 1965), both authors implying that the mechanism is the same as in other guidance responses. However, there is only a superficial resemblance between this bidirectional behaviour, imposed by confinement to a linear region of the substratum, and the bidirectional behaviour of a 'true' guidance which is a response to an anisotropic tensor field and where, in a uniform field, the guidance response is independent of the cell's position in the environment. I do not wish to imply that guidance due to the parallel adhesive boundary response is any less important than guidance due to a tensor field as a possible mechanism for directing cell behaviour *in vivo,* but rather to illustrate the difference between the mechanisms. Indeed, this difference may prove not to be very fundamental since very many closely spaced strips of alternating adhesiveness may approximate to a uniform field and may turn out to elicit guidance. Ohara & Buck (1979) have proposed a mechanism where the confinement of the elongate focal adhesions of fibroblasts to very narrow strips of the substratum may impose a bidirectional behaviour. My preliminary experiments (unpublished) with very close parallel strips of alternating adhesiveness have so far failed to support this proposal.

As part of a study on the guidance of fibroblasts by substratum curvature, Dunn & Heath (1976) have investigated an example of a response to a boundary between two isotropic tensor fields. In each case the boundary was simply a sharp convex change in inclination in an otherwise plane substratum. Changes in inclination of 8° or more often retarded or stopped the locomotion of cells meeting the boundary; this was usually accompanied by a deflection in the direction of locomotion which often resulted in a cell following the boundary. This response differs from the adhesive boundary response in that it is independent of the side from which the boundary is approached. As with the vector boundary, two closely spaced parallel tensor boundaries are probably capable of restricting cell locomotion to two directions; a series of alternating convex and concave parallel boundaries approximates to a uniform field and the finely grooved or ruled substrata used by Weiss (1959) and Ohara & Buck (1979) elicit a guidance response.

Conclusion

The three basic types of response of a cell to its environment – the simple kinesis, the taxis and the guidance – apparently correspond to properties of the environment which are described by quantities having an ascending order of complexity: scalar, vector and tensor quantities. In environments which are uniform with respect to a given quantity, only vector and tensor quantities may contain directional information and, of uniform environments, only vector and tensor fields are capable of eliciting directional cell behaviour. This consideration of the information content of the environment is not a new approach to the description of behaviour patterns but is often implicit in the literature. For example, a type of taxis behaviour has been reported (the 'transverse reactions' of Fraenkel & Gunn, 1940, which have been included in the term 'taxis' by Patlak, 1953*b*) in which locomotion is at right angles to the direction of the vector field. In two dimensions this yields a bidirectional behaviour which is indistinguishable from the behaviour due to a guidance and in three dimensions it is indistinguishable from what I have called an oblate guidance. It was recognized as a taxis not from the behaviour pattern but because it takes place in a vector field.

By following this reasoning it is clear that a klinokinesis with perfect adaptation should be called a temporal taxis (as indeed it now commonly is) since the response is due to the vector information in the environment and not to the scalar information. In fact, a simple model of a temporal taxis need not differ very much from a simple model of a spatial taxis; in the one case, a single receptor samples the environment in two or more places by moving between them as a result of cell locomotion whereas the spatial taxis may have two or more receptors, fixed relative to the cell, which sample simultaneously. As I have suggested, the principle of a moving receptor with memory may also be applied to a spatial taxis, provided that its movement is not derived entirely from the displacement of the cell.

The three types of physical quantity I have discussed are not the only ones but they are the first three ranks of tensor quantities. What I have called a tensor should properly be called a 'tensor of the second rank'; a vector is a tensor of the first rank and a scalar is a tensor of zero rank. It is possible to extend the ideas of a tensor to ranks higher than two. A complete description of the elastic properties of a material, that is the relationship between stress and strain, requires a tensor of the fourth rank and it is not inconceivable that cells may respond to such properties of their environment.

Unidirectional behaviour need not be exclusively the result of taxis mechanisms. Purely random behaviour (type 1) in a closed system can result in a small but real unidirectional component if one end of the system is initially more dense than the other. This is, to some extent, analogous to molecular diffusion but, with most metazoan cells, the effect is augmented by contact inhibition of locomotion which results in a tendency of cells to disperse from regions of higher density (Abercrombie & Heaysman, 1954). If the cells are also in a guidance field which is appropriately oriented, the unidirectional component may become very marked and there may be yet a further contributing factor: in a strong two-dimensional guidance field or a strong three-dimensional prolate guidance field, changes in direction of the cells are restricted to 180° and such reversals of locomotion may have a low probability of occurrence. I have mentioned that a simple kinesis may also result in behaviour with a small unidirectional component if the cell distribution has not reached equilibrium. With an extreme simple kinesis, where the cell motility diminishes to zero in one region of the environment, equilibrium is only reached once the cells are all 'trapped' in that region. In combination with a guidance this may be a very effective mechanism for moving cells from one part of the embryo to another during development.

These instances of cell behaviour resulting from composite mechanisms serve to illustrate the fallibility of assigning an observed behaviour, in the absence of thorough analysis, to a specific mechanism. They also suggest the large range and complexity of behaviour which is possible with such simple mechanisms in combination with each other. Such combinations of simple mechanisms may even account for the enormous complexity of cell behaviour during embryonic development where both the cell environment and the responsiveness of individual cells may change continuously and where populations of cells may be heterogeneous in their responsiveness.

Acknowledgements are due to Michael Abercrombie, Charles Jardine and Sally Zigmond, with whom I have frequently discussed the topics presented here, and to Adam Middleton for his helpful criticism of the manuscript. I am at present supported by a Medical Research Council project grant.

References

Abercrombie, M. & Heaysman, J. E. M. (1954). Observations on the social behaviour of cells in tissue culture II 'Monolayering' of fibroblasts. *Experimental Cell Research,* **6,** 293–306.

Albrecht-Buehler, G. (1977). Daughter 3T3 cells: are they mirror images of each other? *Journal of Cell Biology,* **72,** 595–603.

Albrecht-Buehler, G. (1979*a*). Navigation of nonneuronal cells. In *The Role of Intercellular Signals: Navigation, Encounter, Outcome,* ed. J. G. Nicholls, pp. 75–96. Weinheim: Verlag Chemie.

Albrecht-Buehler, G. (1979*b*). The angular distribution of directional changes of guided 3T3 cells. *Journal of Cell Biology,* **80,** 53–60.

Blakemore, R. P., Frankel, R. B. & Kalmijn, Ad. J. (1980). South-seeking magnetotactic bacteria in the Southern Hemisphere. *Nature, London,* **286,** 384–5.

Bursell, E. & Ewer, D. W. (1950). On the reaction to humidity of *Peripatopsis moseleyi* (Wood-Mason). *Journal of Experimental Biology,* **26,** 335–53.

Carter, S. B. (1965). Principles of cell motility: the direction of cell movement and cancer invasion. *Nature, London,* **208,** 1183–7.

Curtis, A. S. G. & Varde, M. (1964). Control of cell behaviour: topological factors. *Journal of the National Cancer Institute,* **33,** 15–26.

Dunn, G. A. (1980). Mechanisms of fibroblast locomotion. In *Cell Adhesion and Motility,* ed. A. S. G. Curtis & J. D. Pitts, pp. 409–23. Cambridge, England: Cambridge University Press.

Dunn, G. A. & Ebendal, T. (1978). Contact guidance on oriented collagen gels. *Experimental Cell Research,* **111,** 475–79.

Dunn, G. A. & Heath, J. P. (1976). A new hypothesis of contact guidance in tissue cells. *Experimental Cell Research,* **101,** 1–14.

Elsdale, T. & Bard, J. (1972). Collagen substrata for studies on cell behaviour. *Journal of Cell Biology,* **54,** 626–37.

Ewer, D. W. & Bursell, E. (1950). A note on the classification of elementary behaviour patterns. *Behaviour,* **3,** 40–7.

Feller, W. (1966). *An Introduction to Probability Theory and Its Applications,* vol. **2.** New York: John Wiley.

Feynman, R. P., Leighton, R. B. & Sands, M. (1963). *The Feynman Lectures on Physics*. Reading, Massachusetts: Addison-Wesley.

Fraenkel, G. S. & Gunn, D. L. (1940). *The Orientation of Animals: Kineses, Taxes and Compass Reactions*. Oxford: Clarendon Press.

Gail, M. H. & Boone, C. W. (1970). The locomotion of mouse fibroblasts in tissue culture. *Biophysical Journal,* **10,** 980–93.

Gail, M. H. & Boone, C. W. (1972). Cell-substrate adhesivity: a determinant of cell motility. *Experimental Cell Research,* **70,** 33–40.

Gunn, D. L. (1975). The meaning of the term 'klinokinesis'. *Animal Behaviour,* **23,** 409–12.

Harris, A. (1973). Behaviour of cultured cells on substrata of variable adhesiveness. *Experimental Cell Research,* **77,** 285–97.

Harrison, R. G. (1912). The cultivation of tissues in extraneous media as a method of morphogenetic study. *Anatomical Record,* **6,** 181–93.

Holmes, S. J. (1903). Phototaxis in *Volvox*. *Biological Bulletin. Marine Biological Laboratory, Woods Hole, Mass.,* **4,** 319–26.

Jennings, H. S. (1906). *Behaviour of the Lower Organisms*. New York: Macmillan Co.

Keller, H. U., Wilkinson, P. C., Abercrombie, M., Becker, E. L., Hirsch, J. G., Miller, M. E., Ramsey, W. S. & Zigmond, S. H. (1977). A proposal for the definition of terms related to locomotion of leucocytes and other cells. *Clinical and Experimental Immunology,* **27,** 377–80.

Keller, H. U., Wissler, J. H., Hess, M. W. & Cottier, H. (1978). Distinct chemokinetic and chemotactic responses in neutrophil granulocytes. *European Journal of Immunology,* **8,** 1–7.

Macnab, R. M. & Koshland, D. E. (1972). The gradient sensing mechanism in bacterial chemotaxis. *Proceedings of the National Academy of Scienes of the U.S.A.,* **69,** 2509–12.

Mardia, K. V. (1972). *Statistics of Directional Data*. London and New York: Academic Press.

Ohara, P. T. & Buck, R. C. (1979). Contact guidance *in vitro*. A light, transmission, and scanning electron microscope study. *Experimental Cell Research,* **121,** 235–49.

Patlak, C. S. (1953*a*). Random walk with persistence and external bias. *Bulletin of Mathematical Biophysics,* **15,** 311–38.

Patlak, C. S. (1953*b*). A mathematical contribution to the study of orientation in organisms. *Bulletin of Mathematical Biophysics,* **15,** 431–76.

Roberts, A. M. (1970). Geotaxis in motile micro-organisms. *Journal of Experimental Biology,* **53,** 687–99.

Skellam, J. G. (1951). Random dispersal in theoretical populations. *Biometrika,* **38,** 196–218.

Ullyot, P. (1936). The behaviour of *Dendrocoelum lacteum*. I. Responses at light-and-dark boundaries. II. Responses in non-directional gradients. *Journal of Experimental Biology,* **13,** 253–78.

Weiss, P. (1929). Erzwingung elementarer Strukturverschniedenheiten am *in vitro* wachsenden Gewebe. *Archiv für Entwicklungsmechanik der Organismen,* **116,** 438–544.

Weiss, P. (1959). Cellular dynamics. *Revue of Modern Physics,* **31,** 11–20.

Wilkinson, P. C. & Allan, R. B. (1978). Assay systems for measuring leucocyte locomotion: an overview. In *Leucocyte Chemotaxis,* ed. J. I. Gallin & P. G. Quie, pp. 1–23. New York: Raven Press.

Wilkinson, P. C. & Lackie, J. M. (1979). The adhesion, migration and chemotaxis of leucocytes in inflammation. In *Current Topics in Pathology,* **68,** *Inflammatory Reaction,* ed. H. Z. Movat, pp. 47–88. Berlin: Springer Verlag.

Zigmond, S. H. (1974). Mechanisms of sensing chemical gradients by polymorphonuclear leukocytes. *Nature, London,* **249,** 450–52.

Zigmond, S. H. & Sullivan, S. J. (1979). Sensory adaptation of leukocytes to chemotactic peptides. *Journal of Cell Biology,* **82,** 517–27.

H.-U. KELLER

The relationship between leucocyte adhesion to solid substrata, locomotion, chemokinesis and chemotaxis

Orientation reactions are instrumental in guiding actively moving cells or organisms to specific targets (Fraenkel & Gunn, 1940). They are usually associated with activities which are vital for the survival of the species such as fertilization, morphogenesis, nutrition and defence reactions. Leucocyte locomotion and chemotaxis are supposed to serve in the defence against microorganisms. Defects in locomotion and the chemotactic response are associated with increased susceptibility to infectious disease (Miller, 1975; Quie & Cates, 1978). The immediate accumulation of granulocytes in the site of infection seems to be essential for resistance. The chance that inoculation of bacteria will produce infectious disease is drastically increased if emigration of neutrophils to the site of infection is delayed by a few hours (Miles, Miles & Burke, 1957). Furthermore, chemotaxis of granulocytes and monocytes may also play an important role in repair mechanisms such as the removal of breakdown products of tissues and in the digestion of fibrin (Bessis & Burté, 1965; Kay & Kaplan, 1975).

Considerable numbers of granulocytes can be mobilized to specific sites *in vivo*. We have calculated that about 100 000 cells per second migrate into the peritoneal cavity of rabbits during the first three hours following intraperitoneal injection of casein. At the peak of the response the immigration rate is considerably higher. The fact that leucocytes are accumulating specifically at the site of injury is just one aspect of the more general notion that inflammatory responses are in essence localized processes. As far as leucocytes are concerned this may have several advantages if one chooses to take a teleological point of view. The response becomes more economic without losing its efficiency in antimicrobial defence. Furthermore, other effects which are considered to be undesirable, for instance necrosis following excessive release of lysosomal enzymes, become at least localized. Both aspects, defective defence mechanisms or unwanted tissue damage, are of major concern to clinical investigators.

For several reasons leucocytes also represent a fascinating model for studying various aspects of cell locomotion. Granulocytes seem to move much faster than most other metazoan cells. In contrast to the latter they are undoubtably capable of responding chemotactically (McCutcheon, 1946; Harris, 1961; Zigmond,

1977). It is also of general interest that they lack contact inhibition of locomotion (Armstrong & Lackie, 1975). Normal leucocytes show a marked propensity for active locomotion provided they can make adequate contact with a suitable substratum; this happens when the cells marginate and stick to the vascular endothelium *in vivo*. The view that leucocytes emigrate into the tissues in response to chemotactic agents released from the site of injury is widely accepted. There is, however, very little direct proof for this (Keller & Sorkin, 1968), but indirect evidence is steadily growing. Studies on the effects of cytotaxins (substances which will elicit a chemotactic response) *in vivo,* on deficiencies of chemotaxis, and on the effects of pharmacological agents strongly support the supposition that chemotaxis is essential for efficient leucocyte accumulation *in vivo*. Furthermore, it has been demonstrated that granulocytes and monocytes can exhibit marked chemotactic orientation *in vivo,* resulting in directional locomotion of cells towards the target (McCutcheon, Wartman & Dixon, 1934; Harris, 1961; Bessie & Burté, 1965; Zigmond, 1977).

In summary, the available data suggest that contact with the substratum, locomotion and chemotactic orientation are involved in emigration and attraction of leucocytes to sites of inflammation and tissue damage. Furthermore, there is evidence for a complex and intimate relationship between these and other leucocyte functions such as release of lysosomal enzymes, aggregation, neutropenia and the respiratory burst. On the one hand, adherence to blood vessels is a prerequisite for leucocyte emigration which is in turn required for phagocytosis and intracellular killing of bacteria in the tissue. On the other hand, phagocytosis and degranulation result in release of various mediators such as chemokinetic factors, cytotaxins and chemotactic-factor inactivators (Keller, Hess & Cottier, 1975; Goetzl & Austen, 1974; Clark & Klebanoff, 1979; Brozna, Senior, Kreutzer & Ward, 1977).

Several investigators have been struck by the fact that chemotactic formylmethionyl peptides stimulate multiple functions, probably by a common pathway. They found no cases where the biological responses to a given chemotactic factor could be dissociated (Kreutzer *et al.*, 1978; Becker, 1979). Others have presented evidence to suggest that at least some functions can be dissociated more or less from one another (Keller, Wissler, Hess & Cottier, 1978*b*; Cramer & Gallin, 1979; Marasco, Becker & Oliver, 1980). This point is of general interest in leucocyte physiology. It seems important to know whether locomotion, chemokinesis, chemotaxis and other functions are mechanistically connected to form an inseparable complex or whether, and to what extent, these functions can be dissociated from one another. It might for instance be quite difficult to investigate which biochemical reactions, if any, are specifically involved in a particular leucocyte response, e.g. chemotaxis, as long as this point has not been clarified. The present review is mainly concerned with the question as to whether there is a strict relationship between adhesion to solid substrata, chemokinesis and chemotaxis, respectively. Various forms of a relationship between different leucocyte functions have to be considered.

(1) Each response is an integral part of phagocyte activation in general and chemicals, in particular cytotaxins, interacting with a multifunctional receptor trigger all these functions. None of the responses can be dissociated from any of the others.

(2) The functions can be partially or wholly dissociated with respect to the initiating stimulus and/or intracellular pathways. This could be due to a variety of causes. Responses may occur in a differential way because the type of receptors and/or receptor occupancy or the intracellular pathways for transducing and/or effecting the response are partially or wholly different. Mediators may or may not have polyvalent biological activity because they have different active groups on the same molecule. It is also conceivable that the induction of a response requires more than one triggering factor in such a way that a distinct pattern of stimulation is needed to trigger one particular function. Synergistic and facilitating effects may be mediated by secondary messengers.

(3) The responses may be linked sequentially by means of extracellular mediators rather than by common intracellular pathways. Phagocytic, chemotactic or other stimuli may for instance induce release of lysosomal enzymes and other mediators which could in turn modulate or initiate activities in the cell from which they have been released or in other leucocytes.

Terminology and concepts of leucocyte behaviour

Concepts for orientation reactions and the corresponding terminology largely arose in the second half of the nineteenth and during the early twentieth century. This historical development is still of interest because some of the confusion which persists in terminology reflects the diverse sources of these concepts. Furthermore, diverging definitions often imply that biologists are holding different *a priori* views. Their approach to the problem varies accordingly. The use of the terms chemotaxis and chemokinesis exemplifies this point.

Pfeffer proposed the term chemotaxis (chymiotaxis) in 1884 to describe the morphological orientation of fern sperms toward the eggs which release malic acid. In Pfeffer's view (1884, 1904) the term chemotaxis was basically equivalent to tropism, but for historical reasons tropism was used for organisms which are unable to locomote whereas chemotaxis was associated with orientation reactions of locomoting cells. Pfeffer observed that malic acid determined the morphological orientation of sperms without changing their speed. He also noted that some chemicals influenced the speed as well as the orientation of other organisms but he considered this to be a coincidence rather than a strict correlation. He therefore insisted that the term chemotaxis should be restricted to orientation reactions whereas changes in speed were, at least in his view, no essential part of the reaction. He pointed out that changes in speed might make it difficult if not impossible to quantitate chemotaxis by measuring cell accumulation.

This warning was largely disregarded by many workers studying leucocyte chemotaxis by measuring cell accumulation. We were obviously misled by the fact that chemotaxis is by tradition associated with cells or organisms capable of active locomotion. Taking locomotion for granted, cell accumulation in response to chemical agents has been used as a measure for chemotaxis. In contrast to Pfeffer's view, this concept precludes any reasonable differentiation between mere locomotor responses and orientation. Under these conditions even starting and stopping reactions may be mistaken for chemotactic responses.

Rothert (1901) who worked for a short time in Pfeffer's laboratory, observed that some but not all chemotactic agents also produced immobilization of otherwise motile spores. Therefore he coined the term chemokinesis to distinguish chemically induced changes in the rate of locomotion (increase or decrease) from morphological orientation (chemotaxis).

A certain confusion arose because Rothert (1901) discovered a third type of response which he considered to be different from his description of chemokinesis and Pfeffer's chemotaxis. He observed that bacteria accumulated in a chemical gradient without becoming morphologically oriented to any extent. His observations suggested the bacteria were induced to turn when moving away from the highest concentration. Rothert as well as Pfeffer insisted that the phenomenon was different from Pfeffer's chemotaxis and he and Pfeffer called it 'apobatische Taxis' (Rothert, 1901) or 'phobotaxis' (Pfeffer, 1904). Investigators studying bacteria continue to use the term chemotaxis to describe the form of behaviour which leads to accumulation of these cells in chemical gradients. According to Gunn (Gunn, Kennedy & Pielou, 1937; Fraenkel & Gunn, 1940) this form of response, in which the rate of random turning is determined by the intensity of stimulation, is called klinokinesis.

A recent proposal for the definition of terms related to locomotion of leucocytes and other cells (Keller *et al.*, 1977*b*) is fairly close to the system of Fraenkel & Gunn (1940). There is, however, a significant difference. The system suggested by Fraenkel & Gunn (1940) was devised to describe and explain orientation or at least accumulation of locomoting cells or organisms. These investigators took little interest in locomotor reactions unless they were instrumental in orientation. The recent proposal for terms related to locomotion of leucocytes and other cells is more comprehensive: it can also be used to describe changes of random locomotion in the absence of a gradient, including responses which lead to dispersal rather than to accumulation of cells. This difference has an impact on the experimental approach. Whereas Fraenkel & Gunn (1940) postulate that orthokinesis as well as klinokinesis can only be demonstrated in a gradient, the other concept implies that chemokinesis can also be observed in the absence of a gradient. Under these conditions chemokinetic responses maintain or induce the random distribution of cells or organisms. Studies in leucocytes (Keller & Sorkin, 1966; Keller, Hess & Cottier, 1977*a*; Wilkinson & Allan, 1978) demonstrated that orthokinesis can indeed be observed in the absence of a gradient. Klinokinesis has not as yet been demonstrated in leucocytes, but it may well be that this point has not been investigated thoroughly enough. In a gradient, chemotaxis offers more positional information and it is therefore a much more precise guiding principle than klinokinesis.

Regardless of the terminology, it is important to keep in mind that the behaviour of bacteria in a gradient of an 'attractant' is different from classical chemotaxis, as observed in leucocytes, and from orthokinesis. The terms adhesion and attachment have been used in this paper in an operational way. They imply the existence of optically or functionally detectable contact(s) with a substratum.

Locomotion

Locomoting leucocytes are more or less polarized (Lewis, 1934; De Bruyn, 1946; Fukushima *et al.*, 1954). The front is characterized by a lamellipodium consisting of clear hyaloplasm. Granules, which do not normally enter the lamellipodium, are found predominantly in the middle and rear parts of the cells surrounding the nucleus. A tail-like structure (uropod) is usually found at the rear end of the cells. This polarization is exaggerated in a grotesque way if cells are briefly heated at 46°C and even more if they are chemotactically stimulated under these conditions. Finally the leading lamellipodium may become detached and move away on its own. These observations indicate that the locomotor machinery in the front part of the cell, in particular the leading lamellipodium, is essential for locomotion and chemotaxis, whereas the rear end including nucleus and granules are not needed (Keller & Bessis, 1975).

Whereas the cytoplasm of the locomoting cell is flowing forward, the movement on the membrane appears to be backwards. Lateral lamellipodia move backwards towards the tail (Ramsey, 1972*b*). Furthermore, surface-bound lectins are transported to the rear end of the cells (Ryan, Borysenko & Karnovsky, 1974). Lewis (1934) and De Bruyn (1946, 1947) described some characteristic features of leucocyte locomotion. They observed the outline of moving cells to be undulant with areas where the body of the cells is narrow (so-called constriction rings) alternating with lateral protuberances. Both features are fairly stationary with respect to the substratum. Granules in the cytoplasm were seen to be streaming forward through the constriction rings. The findings were interpreted to mean that leucocytes were being moved forward by cytoplasmic flows generated in the rear end of the cell. Differences in the flow rate and the Brownian movement of cytoplasmic granules were considered as evidence of gel⇆sol changes. The hypothesis that gel⇆sol transformation plays an essential role in the mechanism of amoeboid movement has been prevalent for decades (De Bruyn, 1947), but direct proof is still lacking (Ramsey, 1972*b*). Nowadays the transformation of microfilaments into a diffuse mesh and vice versa is sometimes considered as the equivalent of this sol⇄gel transformation (Schiffmann & Gallin, 1979). A diffuse meshwork has been observed in cytoplasmic fragments capable of locomotion (Keller, unpublished results). The contractile proteins account for about 15% of the total protein. More than 10% of the total cell protein is actin. Actin-binding-protein and myosin have also been demonstrated in leucocytes (Stossel & Hartwig, 1976; Stossel, 1978). It seems obvious that these contractile proteins are involved in locomotion. The association of defective leucocyte locomotion and impaired actin polymerization confirms this view (Boxer, Hedley-Whyte, Glader & Stossel, 1974). Locomotion requires energy which is generated via the glycolytic pathway (Carruthers, 1966; Zigmond & Hirsch, 1972) and the locomotor activity is optimal at body temperature (Dahlgren, Hed, Magnusson & Sundqvist, 1979). These findings argue against the hypothesis that leucocytes are exclusively moved by external physical forces such as surface tension or adhesion to the substratum (Berthold, 1886; Carter, 1967).

Unlike bacteria, leucocytes do not exhibit a fixed motor pattern in the form of well defined turns. Sometimes one has in fact to define a turn before one can decide whether a leucocyte has made one or not. Leucocytes show random locomotion in the sense that the direction and the amount of turning cannot be predicted. There is an equiprobable distribution of intersegmental angles. The mean square displacement of the cells is proportional to time. This leads to random distribution. But the formation of new lamellipodia of individual leucocytes is nevertheless biased. Lamellipodia giving a new direction tend to be formed in angles less than 90°, with the mean angle of turn being about 50° (Zigmond, 1978). Therefore the path may often look relatively persistent, at least for short distances. Radical changes of direction are less frequent and usually occur after the cell has rounded up for a moment.

An average speed of 6.5 μm/min was measured *in vivo*. The fastest cells in tissues move at a rate of 11.8 μm/min (Cliff, 1966). The speed measured *in vitro* was found to vary between 0 and 40 μm/min depending on the test conditions (De Bruyn, 1945; Keller, Barandun, Kistler & Ploem, 1979*a*). Variations in the speed of individual cells from the same batch is quite considerable *in vivo* and *in vitro*. In a population moving at an average speed of approximately 10 μm/min, it varied between 0 and 22 μm/min (Ramsey, 1972*a*).

Locomotion and adhesion

Neutrophil precursors proliferate in the bone marrow. The deformability of the cells is probably instrumental in regulating the passage through the endothelium separating the bone marrow from the circulating blood (Lichtman, 1970). Intact cells travel passively in the circulating blood until they join the marginated pool by adhering to the wall of small vessels and/or emigrating into the tissues. Thus attachment to the endothelium of small vessels marks the transition from passive transport to active locomotion. It is difficult to determine the shape of circulating leucocytes *in vivo* but it seems reasonable to assume that they are more or less spherical. Leucocytes spread and develop locomotor activity following contact with suitable substrata such as the vascular endothelium in the process of margination. There are also studies *in vitro* which show that leucocytes are round and immobile immediately after the preparation has been set up but spread and start locomotion afterwards (Zigmond & Hirsch, 1973). This seems to fit into the general view that the shape of cells is maintained by attachment to the substratum (Abercrombie, 1980). A relationship between locomotion, shape and adhesion has also been found to some extent in leucocytes (Keller *et al.*, 1978*b*). De Bruyn (1946) suggested that constriction rings are caused by external factors such as obstacles on the coverglass. Rounding-up of a cell often indicates detachment.

Observations of this kind that have been made in various cell types have had a considerable impact on the theories of cell locomotion. One theory implies that shape changes including pseudopod formation and locomotion are exclusively generated and maintained by adhesive forces between cells and the substratum. Locomotion takes place whenever adhesion to the substratum is stronger at the front than at the rear end of the cell (Carter, 1967).

So far, there is no convincing evidence that this theory can be applied to leucocyte locomotion. In some situations the tail adheres even more strongly than the advancing front of the cell (Keller, Wissler & Ploem, 1979*c;* Keller *et al.*, 1979*a*). Furthermore, the formation of pseudopods has been observed in leucocytes that did not attach or locomote to any measurable extent (Keller *et al.*, 1979*a*) and it was found to continue in cells that had detached from the substratum (Armstrong & Lackie, 1975). Studies on granulocyte migration through endothelium in culture suggest that adhesion of cells to the endothelium and migration through it are separately controlled (Beesley *et al.*, 1979). The finding that leucocyte locomotion is highly temperature-dependent and requires metabolic energy further supports the impression that leucocyte locomotion is not exclusively determined by external physical forces. Intrinsic cellular mechanisms seem to play an essential role as well.

Alternative models have been developed to explain cell locomotion. They take into account that locomotion may be an expression of the intrinsic motility of leucocytes. These concepts nevertheless imply that adhesion to the substratum plays a considerable role in one way or another. Dellinger and Mast (for review see De Bruyn, 1947) believed that contact with the substratum was limited to a few areas which could function as feet. Recent studies on neutrophils showed that the lamellipodia which locomoting neutrophils protrude may be in constant intimate contact with the substratum or show looping movements without such continuous contact (Armstrong & Lackie, 1975). The attachment of leucocytes can vary in intensity and with respect to size and localization of the area of contact. Neutrophils sometimes exhibit two or more such areas, for instance one underneath the advancing lamellipodium, the other(s) underneath the body and/or the tail. Neutrophils with a single grey area of contact underneath the middle and/or front part of the polarized cell were also observed by interference-reflection microscopy (Keller *et al.*, 1979*a*). Focal adhesions of the type found in fibroblasts can be totally absent in efficiently moving neutrophils. Therefore, it appears unlikely that they represent the kind of feet required for neutrophil locomotion (Armstrong & Lackie, 1975). If neutrophils walk on foot-like processes, these are probably beyond the resolving power of interference-reflection (reflection-contrast) microscopy. Izzard & Lochner (1976) suggested that the grey areas of contact appearing in interference-reflection microscopy correspond to a separation of approximately 100 nm. But it has not as yet been convincingly excluded that minute black areas of adhesion in some situations produce grey or black pictures depending on the density of these sites – which would be equivalent to very small feet. Alternatively, it is conceivable that such feet are not needed at all. The cell could move forward because the cell membrane, including the areas adhering to the substratum, is moving backwards.

Efficiently moving leucocytes exhibit grey areas of contact in interference-reflection microscopy (Armstrong & Lackie, 1975; Keller *et al.*, 1979*a*). One may nevertheless raise the question whether this form of contact is under all circumstances required for locomotion. Neutrophils in 1% fibrinogen show no substantial adhesion to glass or to mixed esters of cellulose and they produce no grey or black areas of contact. They show little or no locomotion on glass, but efficient translocation in the three-dimensional network of micropore filters (Keller *et al.*, 1979*a;* Keller *et al.*, 1980; Lackie & Smith, 1980). One possible explanation is that the intimate contact visualized by grey areas of contact is only necessary on a plane surface. A three-dimensional network gives the cell additional means of translating motility into locomotion. This concept seems reasonable because crawling-like movements can be observed in the absence of detectable adhesion to a solid substratum (Keller & Cottier, 1981). Alternatively one has to consider that the cells may develop different forms of contact in a three-dimensional structure. This point is of some interest in relation to leucocyte locomotion *in vivo*.

Leucocytes advance even at the price of losing substantial amounts of cellular material. Reflection-contrast studies show that locomoting cells may leave parts of the membrane behind if attachment to the substratum is too strong. If adhesion is weak, locomotion is faster and no such material is detectable (Keller *et al.*, 1979*a;* Fig. 2). It is not known whether loss of cellular material in the course of locomotion also occurs *in vivo*.

Chemokinesis

Chemokinetic reactions are characterized by the fact that the speed or frequency of locomotion and/or the frequency and magnitude of turning is determined by chemicals in the environment (Keller *et al.*, 1977*b*). In short, the rate of translocation by active locomotion is increased or decreased without the help of tactic stimuli. It is important to realize that the definition does not presuppose an absolute level of locomotory rate; the response is just relative to a control without the chemical in question. The recorded effects of a chemical can vary with the type of control conditions chosen. The response may depend on whether the chemical is tested on immobile or efficiently locomoting cells, whether the substratum is plane or three-dimensional or whether the chemical in question can bind to the substratum or to the cells, or interact with components of the medium, and so forth. It may therefore be misleading to classify a chemical in absolute terms on the basis of chemokinetic effects measured in a particular test system. It is in fact conceivable that a chemical could exhibit positive or negative chemokinetic properties or be ineffective depending on its concentration and the control conditions. Two essentially different approaches can be envisaged. One is to study the minimal requirements for stimulating locomotion using totally immobilized control cells in a relatively artificial environment such as a glass surface and a simple medium such as plain Gey's solution or saline. This is in itself a most interesting approach but it could well be that such experiments tell us little about the *in vivo* effects of a chemical, for instance a cytotaxin. An alternative approach is to use efficiently locomoting control cells in complex media such as plasma and more physiological surfaces such as endothelial cells in an attempt to mimic *in vivo* conditions as closely as possible. The results might be more representative of the chemokinetic effects of a chemical under physiological conditions, but provide incomplete information about the potential effects of a chemical in other conditions. A weak chemokinetic activity of a chemical, e.g. a cytotaxin, may be obscured by the more potent chemokinetic effects of constituents of the basic medium.

The situation is further complicated by the fact that the mechanisms involved in chemokinesis are very heterogenous. Orthokinesis can be caused by a variety of factors such as changes in adhesion between cells and substratum, or interference with the cellular metabolism or the locomotor apparatus (Keller *et al.*,

1977*b*). This indicates that the mechanisms underlying the chemokinetic response are often rather unspecific. Specific receptors are not necessarily involved in the response, but they can probably play a role in chemokinesis induced by cytotaxins or immunoglobulins. Klinokinesis has not as yet been demonstrated in leucocytes. If it exists, specific receptor-mediated reactions similar to ones demonstrated in bacteria have to be suspected.

Iodoacetate exerts a negative chemokinetic effect by interfering with the cell's metabolism, in particular the glycolytic pathway (Carruthers, 1966). Phosphorylating and non-phosphorylating organophosphorus compounds have a negative chemokinetic effect which is probably due to their detergent properties rather than their properties as enzyme inhibitors (Woodin & Harris, 1973). The mechanism of action of cytochalasin B may be complex. It could act by suppressing glycolysis and/or by interacting with actin-like filaments (Zigmond & Hirsch, 1972; Ramsey & Harris, 1972). Furthermore, procaine, a local anaesthetic which has been reported to reduce the permeability of cell membranes to cations, and various other anaesthetics, inhibit neutrophil locomotion (Wilkinson, 1975; Moudgil, Allan, Russell & Wilkinson, 1977). Agents interfering with microtubules such as colchicine and vinblastine do not inhibit locomotion unless the concentration is excessively high (Ramsey & Harris, 1972). Wilkinson even reported a positive chemokinetic effect of colchicine on lymphoblasts (Russell, Wilkinson, Sless & Parrott, 1975).

Chemokinesis and adhesion

Recent studies have shown that many chemicals including cytotaxins exert chemokinetic effects by modifying adhesion to the substratum. The degree of correlation between locomotion and adherence depends on the technique used to determine adhesion to solid substrata. We find a better correlation with reflection-contrast microscopy compared to detachment of cells from glass by external force such as shaking (Keller *et al.*, 1979*a*). It matters also whether the whole population or only locomoting cells are analysed. This is probably related to the finding that chemokinetic factors change not only the rate of locomotion but also the proportion of locomoting cells (Keller *et al.*, 1977*a;* Lackie & Smith, 1980). Reflection-contrast microscopy provides information on the size of areas of adhesion and, perhaps more importantly, on the intensity of adhesion. Neutrophils move most efficiently on a plane surface if adhesion is intermediate as shown by grey areas of adhesion. Large black areas indicating very close contact are associated with spreading, anchoring and immobilization of neutrophils. Locomotion is also absent if the cells fail to make contact with the substratum (Keller *et al.*, 1979*a;* Lackie & Smith, 1980). The positive orthokinetic effect of low-molecular-weight factors and of proteins such as human serum albumin or fibrinogen correlates with the capacity of these chemicals to decrease the adhesion of cells which would otherwise adhere strongly to the substratum (Keller *et al.*, 1977*a*, 1979*a*; Lackie & Smith, 1980). The effects of γ-globulin preparations seem to be more complex but they are at least partly mediated by changes in adhesion. The speed of movement of the cells as well as the strength of adhesion can be further regulated in a dose-dependent fashion by mixing chemokinetic factors with antagonistic effects. Thus changing adhesion is one of several ways of controlling the speed of locomotion *in vitro* (Keller *et al.*, 1979*a*). It is of interest that a correlation between adhesion and chemokinesis has also been observed with cytotaxins such as f-Met-Leu-Phe and $C5a_{des\ Arg}$ (Fehr & Dahinden, 1979; Keller, Wissler & Damerau, 1981; Damerau & Keller, unpublished). This indicates that at least some cytotaxins can produce chemokinetic effects in a similar way to non-chemotactic chemokinesins.

It remains to be evaluated to what extent these effects observed with artificial substrata *in vitro* are also relevant for physiological substrata *in vivo*. This problem has been discussed in detail by Lackie & Smith (1980). They suggest that adhesion to endothelial cells *in vitro* is of comparable strength to adhesion to serum-coated glass. It seems possible nevertheless that adhesion can be changed by effects on the endothelial cell surface (Pearson *et al.*, 1979). Immobilization of neutrophils by immune complexes may occur in tissues *in vivo*.

Chemotaxis

Chemotaxis is a reaction by which the *direction* of locomotion of cells and organisms is determined by substances in their environment (McCutcheon, 1946). It is expressed in morphological orientation. Therefore, direct demonstration of chemotaxis requires a visual assay. An excellent system has been developed by Zigmond (1977). The degree of chemotactic orientation is concentration-dependent, and more than 90 % of the neutrophils can become oriented in response to a gradient of chemotactic mediators. The chemotropism index of McCutcheon (1946) is also an appropriate measure. Indirect assay systems such as the filter methods (Keller *et al.*, 1978*a*; Zigmond & Hirsch, 1973) are useful routine techniques but require restrictive and careful interpretation (Keller *et al.*, 1977*b*, 1980; Zigmond, 1978). They measure random and directional locomotion through a filter and chemokinetic effects can drastically modify the results. The activity of several chemotactic factors cannot be measured in filter systems unless chemokinetic activity promoting locomotion is also present (Keller *et al.*, 1977*c*; Keller, Hess & Cottier, 1978*a*; Wilkinson & Allan, 1978; Till, Kownatzki, Seitz & Gemsa, 1979). Other chemotactic factors exhibit sufficient chemokinetic activity to promote passage through the filter (Wilkinson & Allan, 1978). Even in the presence of stimulating factors, chemotactic activity may not be expressed in terms of directional locomotion if the test solution also contains potent immobilizing factors (Keller *et al.*, 1980). Furthermore, the results obtained in a gradient are only representative of chemotaxis if the chemotactic factor lacks additional chemokinetic activity. Otherwise the results have to be analysed by calculation (Zigmond & Hirsch, 1973).

Chemotaxis and adhesion

The following forms of a relationship between chemotaxis and adhesion will be considered. (1) Gradients of adhesion could be instrumental in the orientation reaction in the sense that chemotaxis is equivalent to haptotaxis. (2) Cytotaxins may stimulate or decrease adherence of neutrophils to the substratum, in particular to the endothelium. These changes in adhesion may be irrelevant for chemotaxis but result in margination, neutropenia and chemokinetic effects.

Binding of chemotactic factors to solid substrata has been demonstrated and it has been suggested that it is essential for the attracting capacity of chemotactic factors (Dierich, Wilhelmi & Till, 1977). This raises once again the question whether chemotaxis and any other form of locomotion is due to gradients of adhesion (Carter, 1967). Movement, pseudopod formation and shape changes are not exclusively or primarily determined by the interaction of cells with the substratum (see above). Furthermore, there is evidence to suggest that chemotaxis is not a special case of haptotaxis. Gradients of adhesion did not produce orientation of neutrophils. In addition cytotaxins can induce efficient orientation and directional locomotion without changing adhesion to the substratum (Keller *et al.*, 1979*c*; Keller, Wissler & Damerau, 1981). These findings do not exclude the possibility that haptotaxis could be a special case of chemotaxis, but so far, haptotaxis has not been demonstrated in leucocytes.

Many experiments nevertheless suggest that most cytotaxins may modulate adhesion to substrata *in vitro* and *in vivo*. Such changes may be important in several respects. They could serve to regulate the size of the marginating pool of blood leucocytes, to induce attachment to the walls of blood vessels as a first step of leucocyte emigration and/or to regulate the speed of locomotion, in particular to immobilize cells by high cytotaxin concentrations in the site of injury. It has been demonstrated by experiments *in vivo*, that irritants such as casein or culture filtrates of *Escherichia coli* cause sticking of leucocytes to the endothelium (Atherton & Born, 1972). Furthermore, complement activation leads to increased granulocyte adherence and consequently to margination and neutropenia. It has been suggested that these phenomena are mediated by chemotactic factors (Fehr & Jacob, 1977; Kreutzer *et al.*, 1978). The question remains whether margination and stickiness is really strictly related to the chemotactic activity of these compounds. Numerous studies have been performed to assess the effects of highly purified cytotaxins on adhesion to solid substrata *in vitro*. The results appear to be most contradictory. Some investigators report that cytotaxins increase adhesion to various substrata (O'Flaherty, Kreutzer & Ward, 1978; Fehr & Dahinden, 1979), and others find that it can be decreased (Smith,

Lackie & Wilkinson, 1979) or find no change in adhesion at all (Keller *et al.*, 1979*c*). Adhesion may vary with time, but even if such variation is taken into account, the discrepancies remain (Smith *et al.*, 1979; Boxer *et al.*, 1979; Fehr & Dahinden, 1979). It appears that the effects of cytotaxins on cell adhesion to the substratum are indeed variable for several reasons. Different cytotaxins produce different effects: they may or may not increase adhesion at chemotactic concentrations under identical test conditions (Keller, Wissler & Damerau, 1981). This may be taken as an indication that the capacity to change adhesiveness in a defined way is not an essential attribute of chemotactic factors. There is also evidence to suggest that the adhesive effects of one and the same cytotaxin at the respective chemotactic concentrations may vary according to the test conditions. It has for instance been reported that f-Met-Leu-Phe can decrease adhesion, be ineffective or produce an increase, depending on time, concentration and probably other differences in the test procedure (Smith *et al.*, 1979; Fehr & Dahinden, 1979; Keller, Wissler & Damerau, 1981). Also $C5a_{des\ Arg}$ may or may not decrease adhesion to the substratum depending on the test conditions (Damerau & Keller, unpublished). It would in fact not be surprising if many of the apparently contradictory findings are due to the different test conditions and techniques used by different investigators. Some investigators study the effect of chemicals on neutrophils which are strongly adherent under control conditions, whereas others measure the effect under conditions which permit only weak contact between control cells and the substratum. It seems that, as with chemokinesis, the effect of chemicals on adhesion is always relative to the particular control conditions (see below), rather than an absolute property. Similarly adhesiveness is not an absolute property of the cell. It depends on the environmental conditions and varies accordingly.

In contrast to the CAT 1.6.1 preparation (Wissler, Sorkin & Stecher, 1974) of chemotactic serum peptides, $C5a_{des\ Arg}$ sometimes, but not consistently, produced an increase in retraction fibres indicating a slight increase in adhesion which has not been detected by other methods. This phenomenon was very pronounced and consistent at chemotactic concentrations of f-Met-Leu-Phe (10^{-8}M). At higher concentration (10^{-5}M) adhesion was further increased (Fig. 2). Large black areas were detected by means of reflection-contrast microscopy and the cells became immobilized (Fig. 1, and Keller, Wissler & Damerau, 1981). The 'focal' type of adhesions formed at intermediate concentration (10^{-8}M) could reflect heterogeneities of the cell surface or the substratum.

Chemokinesis and chemotaxis

Directional locomotion depends on chemokinetic as well as chemotactic influences. Thus, chemokinesis and chemotaxis are functionally closely associated. Two forms of relationship will be invoked.

(1) Chemotaxis could be a special case of stimulated locomotion in the sense that stimulation in a gradient leads to directional locomotion instead of stimulated random locomotion. No distinction between chemokinetic and chemotactic factors might then be possible.

(2) The relationship between chemotaxis and chemokinesis is less strict and takes the form of a functional association or cooperation producing synergistic or antagonistic effect on directional locomotion. If so, it should be possible to dissociate these functions at least partially.

The concept that interaction of chemotactic oligopeptides with a multifunctional receptor leads to stimulation of random as well as directional locomotion, in particular chemotaxis, and of many other neutrophil functions (Becker, 1979) might imply that directional locomotion is just a special case of stimulated locomotion. If chemokinesis and chemotaxis were just two different expressions of one and the same reaction, one would expect that any chemokinetic agents would also have chemotactic activity and vice versa. However, chemokinetic factors such as native human serum albumin (HSA) or fibrinogen lack chemotactic activity. On the other hand, cytotaxins such as CAT 1.6.1 (Keller *et al.*, 1978*a*) $C5a_{des\ Arg}$ (Keller, Wissler & Damerau, 1981), the prostaglandins PGE and PGA (Till *et al.*, 1979), ECF (König, Frickhofen & Tesch, 1978; Weller & Goetzl, 1979) or bacteria (Dixon & McCutcheon, 1936; Ramsey,

Fig. 1. Effect of f-Met-Leu-Phe on random and directional locomotion of human neutrophils in Gey's solution containing 2% human serum albumin (HSA). Migration was assessed with the two-filter-count technique (Keller *et al.*, 1980) and an incubation time of 3 and 1 h for random and directional locomotion, respectively. Filled squares, no gradient; filled triangles with dashed line, positive gradient.

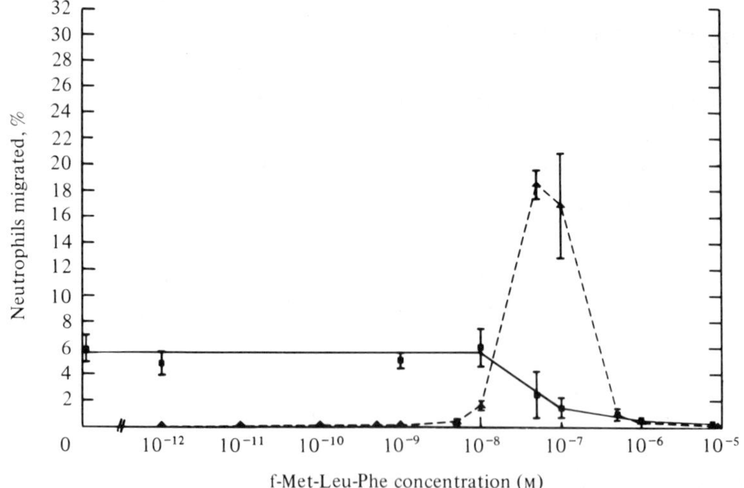

Fig. 2. Phase-contrast (*a*) and reflection-contrast (*b*) pictures of human blood leucocytes on glass in: i, Gey's solution containing 2% HSA; ii, Gey's solution containing 2% HSA and f-Met-Leu-Phe (10^{-8}M); and iii, Gey's solution containing 2% HSA and f-Met-Leu-Phe (10^{-5}M). X735.

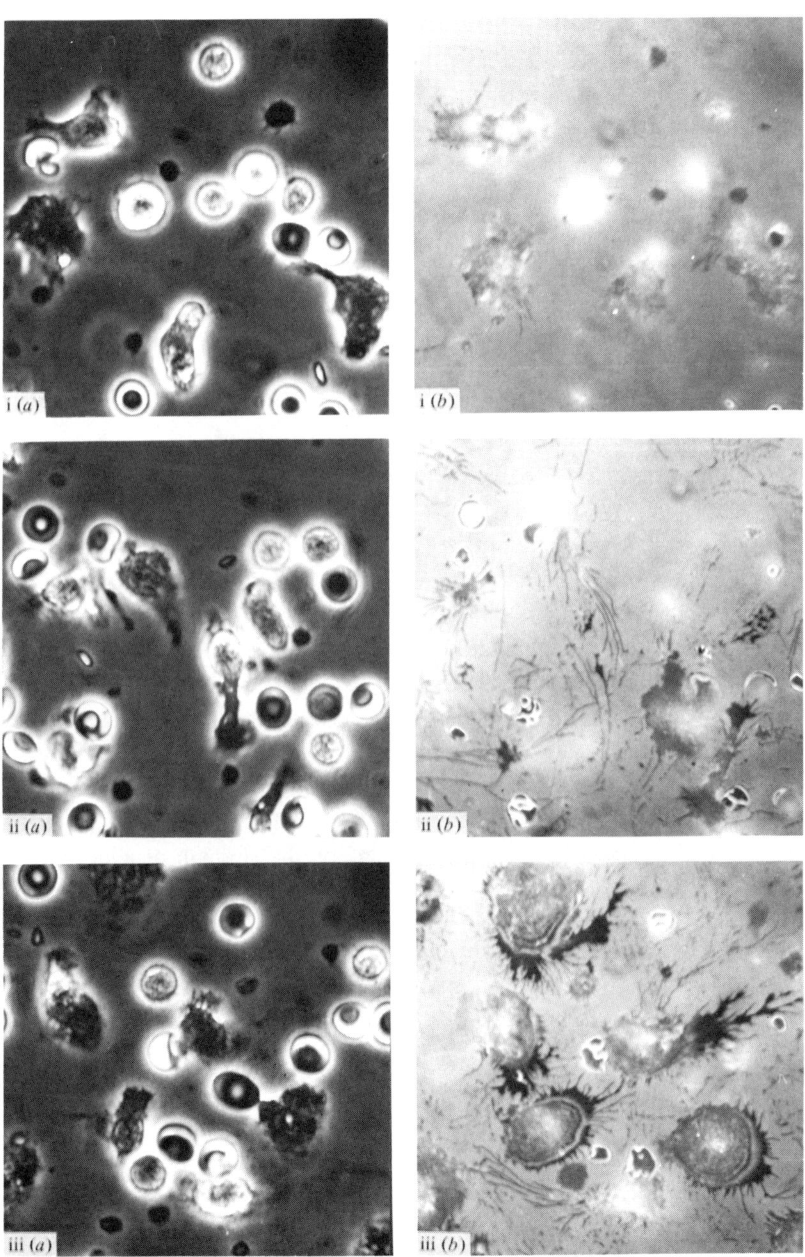

1972*a*) can be chemotactic without exerting significant chemokinetic activity with or without a gradient. Furthermore, the rate of locomotion is not changed following partial deactivation of neutrophils by chemotactic CAT 1.6.1 (Keller *et al.*, 1978*a*). f-Met-Leu-Phe was found to produce diverging effects. The results presented by Showell *et al.* (1976) suggest that f-Met-Leu-Phe has pronounced positive chemokinetic effects at chemotactic concentrations. These results were basically confirmed by Lackie & Smith (1980) who showed that this effect was proportionally much greater at lower bovine serum albumin (BSA) concentrations. At even higher concentration of HSA chemokinesis was indifferent and it actually became negative at high but chemotactic concentrations of f-Met-Leu-Phe (Fig. 1). Furthermore, f-Met-Leu-Phe and $C5a_{des\ Arg}$ differ with respect to their chemokinetic effects under identical test conditions (Keller, Wissler & Damerau, 1981). Differential effects on chemotaxis and the rate of locomotion have also been produced by labelling of plasma membrane with impermeant covalent reagents (Goetzl & Hoe, 1979).

Factors with chemotactic and chemokinetic activity, respectively, can also produce different biochemical effects. Cytotaxins produce a short transient rise in intracellular cyclic AMP whereas chemokinetic albumin has no such effect (Keller, Gerisch & Wissler, 1979*b*). Furthermore chemotactic orientation towards f-Met-Leu-Phe does not require exogenous divalent cations, whereas the speed is regulated by addition or removal of Ca^{++} and Mg^{++} (Marasco *et al.*, 1980). It has also been suggested that cations, mostly Ca^{++}, localize beneath the cell membrane of the leading edge following chemotactic stimulation, but not under conditions of chemokinesis (Cramer & Gallin, 1979). In summary, the data suggest that there is no strict correlation between chemokinesis and chemotaxis.

It is nevertheless of interest that several chemotactic factors can produce chemokinetic effects. Results obtained with f-Met-Leu-Phe and $C5a_{des\ Arg}$ show a close inverse correlation between their effects on the strength of adhesion and the average rate of locomotion. Decreased locomotion in response to f-Met-Leu-Phe is associated with increased adhesion. $C5a_{des\ Arg}$ in Gey's solution exerts a dose-dependent increase of locomotion and neutrophil detachment, whereas $C5a_{des\ Arg}$ in Gey's solution containing 2% HSA produced no changes in adhesion or in the rate of locomotion (Damerau & Keller, unpublished results). Thus with respect to this inverse correlation between adhesion and locomotion, the chemokinetic cytotaxins which have been studied so far are no different from chemokinetic factors without chemotactic activity such as albumin or fibrinogen. This indicates that their chemokinetic effects may also be mediated by changes in adhesion. It remains to be investigated whether other cytotaxins produce similar effects.

Several cytotaxins produce no cell accumulation *in vitro* in the absence of cytokinesins such as HSA or fibrinogen. Chemotactic and positive chemokinetic factors have to cooperate in order to permit efficient directional locomotion to occur (Keller, Wissler, Hess & Cottier, 1977*c;* Keller *et al.*, 1978*a*). Those chemotactic factors which have chemokinetic effects as well may stimulate (Wilkinson & Allan, 1978) or inhibit directional locomotion according to whether the chemokinetic effects are positive or negative. The effects of other chemokinetic factors may be potentiated or antagonized. We find for instance that f-Met-Leu-Phe can antagonize the positive chemokinetic effect of HSA. Under these conditions chemokinetic effects become a major determinant of the response (Fig. 1) and it becomes difficult to use the filter system even as an indirect quantitative measure of chemotaxis, although the increment over the controls (up to 250-fold) is strong evidence for a marked chemotactic effect. But directional locomotion rather than mere orientation is the basis for leucocyte accumulation and the filter technique is most useful to determine this more complex phenomenon.

At present we can only speculate on the putative role of the chemokinetic effects of chemotactic factors *in vivo*. Chemokinetic factors such as HSA are physiological constituents of most body fluids. Therefore it appears uncertain whether positive chemokinetic effects of cytotaxins are of much significance *in vivo;* negative chemokinetic effects would presumably be expressed more readily.

The situation in leucocyte chemotaxis is similar to that reported by Pfeffer (1904) in lower organisms. Some cytotaxins modify the speed of locomotion whereas others do not. It is therefore justifiable to apply Pfeffer's concept, which implies strict separation between orientation and changes in the rate of locomotion, to leucocytes. It is of general interest that these findings in leucocytes correspond with results obtained in vertebrates. Tinbergen & Lorenz (1938) showed in geese that taxis and locomotion are controlled by different mechanisms. Therefore, behaviourists distinguish between fixed action patterns like locomotion and processes primarily depending on external information like kinesis and taxis (Lorenz, 1978). We propose that this distinction can also be made in leucocytes. Crawling movements can be considered as equivalent to a fixed action pattern whereas kinesis and taxis result from external information which is superimposed on the fixed motor pattern (Keller & Cottier, 1981).

Selective directional locomotion

There is considerable evidence to suggest that directional locomotion plays a considerable role in neutrophil accumulation *in vivo*. In some lesions (e.g. immediate type hypersensitivity reactions) the exudate consists mostly of eosinophils, in others (e.g. Arthus type reactions), neutrophils are much more numerous, whereas macrophages are predominant in granulomas. Lymphocytes prevail in many virus-induced lesions. The precise mechanisms which cause this diversity are not known. A variety of factors are probably involved, such as differential emigration of cells, their capacity to survive or to proliferate *in situ* and the recirculation of emigrated cells via the lymphatics.

It has been suggested that differential emigration in acute inflammation is due to cell-specific chemotaxis (Keller & Sorkin, 1967). There has been some further support for this concept by experiments *in vitro* and *in vivo* (Kay, Stechschulte & Austen, 1971; Goetzl, Woods & Gorman, 1977; Czarnetzki, Frosch, Vakilzadeh & Panneck, 1980). Wissler *et al.* (1974) developed a concept explaining changes in cell specificity in a binary system of serum-derived peptides. But other factors such as C5a are chemotactic for neutrophils, eosinophils and monocytes (Weller & Goetzl, 1979; Snyderman, Shin & Hausman, 1971). Thus, the concept of cell-specific chemotaxis may explain cell-specific accumulations in some situations but not in others.

The finding that cell adherence and the rate of locomotion vary with the cell type may provide the basis for an alternative explanation. Macrophages and monocytes are more adhesive as judged by black areas of adhesion under conditions where neutrophils show grey images. Our unpublished experiments *in vitro* suggest that such differences can actually induce selective migration. Preferential immobilization of one or the other cell type could explain variations in the responding cell type and correspondingly of the composition of the inflammatory exudate.

Chemotaxis, chemokinesis and other leucocyte responses

It has been reported that cytotaxins can produce quite a number of effects other than orientation. Apart from chemokinesis and changes of adhesion to solid substrata they can cause changes of membrane potential, aggregation, release of lysosomal enzymes, activation of the hexose-monophosphate shunt leading to increased production of hydrogen peroxide and superoxide *in vitro,* and neutropenia *in vivo* (Fehr & Jacob, 1977; Kreutzer *et al.,* 1978; Becker, 1979; Becker, Sigman & Oliver, 1979; Gallin, Gallin, Malech & Cramer, 1978*a;* Naccache, Showell, Becker & Sha'afi, 1977; Weissmann, Smolen & Korchak, 1980). Different degrees of correlation between various functions have been reported. Some investigators (Kreutzer *et al.,* 1978; Becker, 1979) found no cases in which the different biological responses to a given chemotactic factor could be dissociated. However, several studies have shown that neutrophil aggregation, the release of lysosomal enzymes and of endogenous pyrogen, the activation of the metabolic burst and production of superoxide radicals occur at much higher concentrations than are optimal for directional locomotion (Gallin, Wright & Schiffmann, 1978*b;* Simchowitz & Spilberg, 1979; Spilberg *et al.,* 1978; Cramer & Gallin, 1979; Fehr & Dahinden, 1979; Gander & Chandler,

1980). These marked differences in the dose–response curves lead to the conclusion that although the same chemotactic stimulus can cause both secretion and chemotaxis, the number of receptors stimulated and the sequence of events after stimulation may be quite different (Cramer & Gallin, 1979). This could indicate a partial dissociation between chemotaxis and other responses. Furthermore, it has been pointed out that chemotactic factors alone produce little or no superoxide or release of lysosomal enzymes. Surface contact or cytochalasin B is required in addition for substantial stimulation (Henson, Zanolari, Schwartzman & Hong, 1978; Boxer *et al.*, 1979). This may indicate that the response depends on a pattern of stimuli rather than receptor occupancy by chemotactic factors alone. Various mechanisms which could produce such effects have been discussed (Naccache *et al.*, 1977), including the possibility that synergistic effects could occur if both stimuli enhance calcium fluxes (Wilkinson, 1979).

Biochemical studies

Biochemical events occurring in response to chemotactic stimuli have been studied extensively (for reviews see Gallin, 1980; Schiffmann & Gallin, 1979; Naccache *et al.*, 1977; and Weissmann *et al.*, 1980). Fascinating attempts have been made to analyse the sequence of events in the transduction process. These studies have not been integrated into the present analysis because there is no easy answer to the question as to whether the biochemical events following stimulation with so-called chemotactic factor are specifically related to chemotaxis, to chemokinesis, exocytosis and other responses, or indiscriminately to all of them.

Another difficulty in the analysis of the biochemical data consists in the finding that the response to chemotactic and other stimuli is not restricted to direct intracellular consequences of receptor occupancy. It has been demonstrated that chemokinesis and chemotaxis are linked by extracellular mediators with other cellular responses such as phagocytosis and degranulation (Keller *et al.*, 1975; Gallin *et al.*, 1978*b*). Some of these substances released from leucocytes have chemokinetic or chemotactic properties, others are capable of inactivating cytotaxins (Clark & Klebanoff, 1979; Brozna *et al.*, 1977). The extent to which correlation or dissociation between different responses is due to intracellular pathways and/or factors released from the cells, remains to be investigated.

Conclusion

The relationship between different leucocyte activities such as adhesion to solid substrata, locomotion, chemokinesis and chemotaxis is very complex. The results suggest nevertheless that the functions can be dissociated to some extent. Balanced adhesion to the substratum is required for random and directional locomotion but locomotion is not exclusively determined by external forces such as adhesion to the substratum and chemotaxis is not a special case of haptotaxis. Furthermore there is evidence to suggest that chemokinesis and chemotaxis are at least to some extent different phenomena which can be regulated independently by distinct factors. A close and subtle cooperation between adhesion to the substratum, locomotion, chemokinesis, chemotaxis and maybe other responses seems to be needed in order to achieve cell accumulation by directional locomotion.

References

Abercrombie, M. (1980). The crawling movement of metazoan cells. *Proceedings of the Royal Society London* B, **207,** 129–47.

Armstrong, P. B. & Lackie, J. M. (1975). Studies on intercellular invasion in vitro using rabbit peritoneal neutrophil granulocytes (PMNS). I. Role of contact inhibition of locomotion. *Journal of Cell Biology,* **65,** 439–62.

Atherton, A. & Born, G. V. R. (1972). Quantitative investigations of the adhesiveness of circulating polymorphonuclear leucocytes to blood vessel walls. *Journal of Physiology, London,* **222,** 447–74.

Becker, E. L. (1979). A multifunctional receptor on the neutrophil for synthetic chemotactic oligopeptides. *Journal of the Reticuloendothelial Society,* Suppl. **26,** 701–9.

Becker, E. L., Sigman, M. & Oliver, J. M. (1979). Superoxide production induced in rabbit polymorphonuclear leukocytes by synthetic chemotactic peptides and A23187. *American Journal of Pathology,* **95,** 81–97.

Beesley, J. E., Pearson, J. D., Hutchings, A., Carleton, J. S. & Gordon, J. L. (1979). Granulocyte migration through endothelium in culture. *Journal of Cell Science,* **38,** 237–48.

Berthold, G. (1886). *Studien über Protoplasmamechanik,* pp. 85–129. Leipzig: Verlag von Arthur Felix.

Bessis, M. & Burté, B. (1956). Positive and negative chemotaxis as observed after the destruction of a cell by UV or laser microbeams. *Texas Reports on Biology and Medicine,* **23,** 204–12.

Boxer, L., Hedley-Whyte, T., Glader, B. & Stossel, T. (1974). A primary defect in neutrophil motility. *Clinical Research,* **22,** 384.

Boxer, L. A., Yoder, M., Bonsib, S., Schmidt, M., Ho, P., Jersild, R. & Baehner, R. L. (1979). Effects of a chemotactic factor, *N*-formylmethionyl peptide, on adherence, superoxide anion generation, phagocytosis, and microtubule assembly of human polymorphonuclear leukocytes. *Journal of Laboratory and Clinical Medicine,* **93,** 506–14.

Brozna, J. P., Senior, R. M., Kreutzer, D. L. & Ward, P. A. (1977). Chemotactic factor inactivators of human granulocytes. *Journal of Clinical Investigation,* **60,** 1280–8.

Carruthers, B. M. (1966). Leukocyte motility. I. Method for study, normal variation, effect of physical alterations in environment, and effect of iodoacetate. *Canadian Journal of Physiology and Pharmacology,* **44,** 475–85.

Carter, S. B. (1967). Haptotaxis and the mechanism of cell motility. *Nature, London,* **213,** 256–60.

Clark, R. A. & Klebanoff, S. J. (1979). Chemotactic factor inactivation by the myeloperoxidase–hydrogen peroxide–halide system. *Journal of Clinical Investigation,* **64,** 913–20.

Cliff, W. J. (1966). The acute inflammatory reaction in the rabbit ear chamber with particular reference to the phenomenon of leukocytic migration. *Journal of Experimental Medicine,* **124,** 543–56.

Cramer, E. B. & Gallin, J. I. (1979). Localization of submembranous cations to the leading end of human neutrophils during chemotaxis. *Journal of Cell Biology,* **82,** 369–79.

Czarnetzki, B. M., Frosch, P. J., Vakilzadeh, F. & Panneck, W. B. (1980). Effect of neutrophil-derived eosinophil chemotactic factor (ECF) in human and guinea pig skin. *Journal of Investigative Dermatology,* **74,** 109–11.

Dahlgren, C., Hed, J., Magnusson, K.-E. & Sundqvist, T. (1979). Characteristics of individual polymorphonuclear leucocyte motility obtained with a new opto-electronic method. *Scandinavian Journal of Immunology,* **9,** 537–45.

De Bruyn, P. P. H. (1945). The motion of the migrating cells in tissue cultures of lymph nodes. *Anatomical Record, Philadelphia,* **93,** 295–315.

De Bruyn, P. P. H. (1946). The amoeboid movement of the mammalian leukocyte in tissue culture. *Anatomical Record, Philadelphia,* **95,** 177–92.

De Bruyn, P. P. H. (1947). Theories of amoeboid movement. *Quarterly Review of Biology,* **22,** 1–24.

Dierich, M. P., Wilhelmi, D. & Till, G. (1977). Essential role of surface-bound chemoattractant in leukocyte migration. *Nature, London,* **270,** 351–2.

Dixon, H. M. & McCutcheon, M. (1936). Chemotropism of leucocytes in relation to their rate of locomotion. *Proceedings of the Society for Experimental Biology and Medicine,* **34,** 173–6.

Fehr, J. & Dahinden, C. (1979). Modulating influence of chemotactic factor-induced cell adhesiveness on granulocyte function. *Journal of Clinical Investigation,* **64,** 8–16.

Fehr, J. & Jacob, H. S. (1977). In vitro granulocyte adherence and in vivo margination: two associated complement-dependent functions. *Journal of Experimental Medicine,* **146,** 641–52.

Fraenkel, G. S. & Gunn, D. L. (1940). *The Orientation of Animals. Kineses, Taxes and Compass Reactions.* London, New York: Oxford University Press. Reprinted: New York: Dover Publications, 1961.

Fukushima, K., Senda, N., Miura, H., Ishigami, S. & Murakami, Y. (1954). Dynamic pattern in the movement of leucocyte. I. Motile type of neutrophil in healthy human adult – the behaviour of neutrophil in the process of the stimulation. *Medical Journal of Osaka University,* **5,** 1–46.

Gallin, J. I. (1980). The cell biology of leukocyte chemotaxis. In *Handbook of Inflammation,* ed. L. E. Glynn, J. C. Houck & G. Weissmann, vol. **2,** pp. 299–325. Amsterdam-New York-Oxford: Elsevier/North-Holland Biomedical Press.

Gallin, J. I., Gallin, E. K., Malech, H. L. & Cramer, E. B. (1978*a*). Structural and ionic events during leukocyte chemotaxis. In *Leukocyte Chemotaxis,* ed. J. I. Gallin & P. G. Quie, pp. 123–43. New York: Raven Press.

Gallin, J. I., Wright, D. G. & Schiffmann, E. (1978*b*). Role of secretory events in modulating human neutrophil chemotaxis. *Journal of Clinical Investigation,* **62,** 1364–74.

Gander, G. W. & Chandler, W. (1980). Release of endogenous pyrogen from rabbit buffy coat cells in response to a chemotactic agent. In *Thermoregulatory Mechanisms and their Therapeutic Implications, 4th Int. Symp. on the Pharmacology of Thermoregulation, Oxford 1979,* ed. B. Cox, P. Lomax, A. S. Milton & E. Schonbaum, pp. 82–3. Basel: Karger.

Goetzl, E. J. & Austen, K. F. (1974). Active site chemotactic factors and the regulation of the human neutrophil chemotactic response. In *Antibiotics and Chemotherapy,* ed. E. Sorkin, vol. **19,** pp. 218–32. Basel: Karger.

Goetzl, E. J. & Hoe, K. Y. (1979). Chemotactic factor receptors of human PMN leucocytes. I. Effects on migration of labelling plasma membrane determinants with impermeant covalent reagents and inhibition of labelling by chemotactic factors. *Immunology,* **37,** 407–18.

Goetzl, E. J., Woods, J. M. & Gorman, R. R. (1977). Stimulation of human eosinophil and neutrophil polymorphonuclear leukocyte chemotaxis and random migration by 12-L-Hydroxy-5,8,10,14-Eicosatetraenoic Acid. *Journal of Clinical Investigation,* **59,** 179–83.

Gunn, D. L., Kennedy, J. S. & Pielou, D. P. (1937). Classification of taxes and kineses. *Nature, London,* **140,** 1064.

Harris, H. (1961). Chemotaxis and phagocytosis (1961). In *Function of the Blood,* ed. R. G. MacFarlane & A. H. T. Robb-Smith, pp. 463–94. New York & London: Academic Press.

Henson, P. M., Zanolari, B., Schwartzman, N. A. & Hong, S. R. (1978). Intracellular control of human neutrophil secretion. I. C5a-induced stimulus-specific desensitization and the effects of cytochalasin B. *Journal of Immunology,* **121,** 851–5.

Izzard, C. S. & Lochner, L. R. (1976). Cell-to-substrate contacts in living fibroblasts: an interference-reflexion study with an evaluation of the technique. *Journal of Cell Science,* **21,** 129–59.

Kay, A. B. & Kaplan, A. P. (1975). Annotation. Chemotaxis and haemostasis. *British Journal of Haematology,* **31,** 417–21.

Kay, A. B., Stechschulte, D. J. & Austen, K. F. (1971). An eosinophil leukocyte chemotactic factor of anaphylaxis. *Journal of Experimental Medicine,* **133,** 602–19.

Keller, H. U., Barandun, S., Kistler, P. & Ploem, J. S. (1979*a*). Locomotion and adhesion of neutrophil granulocytes: Effects of albumin, fibrinogen and gamma globulins studied by reflection contrast microscopy. *Experimental Cell Research,* **122,** 351–62.

Keller, H. U. & Bessis, M. (1975). Migration and chemotaxis of anucleate cytoplasmic leukocyte fragments. *Nature, London,* **258,** 723–4.

Keller, H. U. & Cottier, H. (1981). Crawling movements and polarization in non-adherent leucocytes. *Cell Biology International Reports* **5,** 3–7.

Keller, H. U., Gerisch, G. & Wissler, J. H. (1979*b*). A transient rise in cyclic AMP levels following chemotactic stimulation of neutrophil granulocytes. *Cell Biology International Reports,* **3,** 759–65.

Keller, H. U., Hess, M. W. & Cottier, H. (1975). Physiology of chemotaxis and random motility. I. + II. In *Neutrophil Physiology and Pathology,* ed. J. R. Humbert, P. A. Miescher & E. R. Jaffe, pp. 45–67. New York: Grune & Stratton.

Keller, H. U., Hess, M. W. & Cottier, H. (1977*a*). The chemokinetic effect of serum albumin. *Experientia,* **33,** 1386–7.

Keller, H. U., Hess, M. W. & Cottier, H. (1978*a*). The assessment of chemokinesis and chemotaxis. *Symposium on Clinical Aspects of the Complement System,* Schloss Hugenpoet, Germany 1976, pp. 31–42. Stuttgart: Thieme.

Keller, H. U. & Sorkin, E. (1966). Studies on chemotaxis. IV. The influence of serum factors on granulocyte locomotion. *Immunology,* **10,** 409–16.

Keller, H. U. & Sorkin, E. (1967). Studies on chemotaxis. VI. Specific chemotaxis in rabbit polymorphonuclear leucocytes and mononuclear cells. *International Archives of Allergy and Applied Immunology,* **31,** 575–86.

Keller, H. U. & Sorkin, E. (1968). Chemotaxis of leucocytes. *Experientia,* **35,** 641–52.

Keller, H. U., Wilkinson, P. C., Abercrombie, M., Becker, E. L. Hirsch, J. G., Miller, M. E., Ramsey, W. S. & Zigmond, S. H. (1977*b*). A proposal for the definition of terms related to locomotion of leucocytes and other cells. *Clinical and Experimental Immunology,* **27,** 377–80.

Keller, H.-U., Wissler, J. H. & Damerau, B. (1981). Diverging effects of chemotactic serum peptides and synthetic f-Met-Len-Phe on neutrophil locomotion and adhesion. *Immunology,* **42,** 379–83.

Keller, H. U., Wissler, J. H., Damerau, B., Hess, M. W. & Cottier, H. (1980). The filter technique for measuring leucocyte locomotion in vitro. Comparison of three modifications. *Journal of Immunological Methods,* **36,** 41–53.

Keller, H. U., Wissler, J. H., Hess, M. W. & Cottier, H. (1978*b*). Distinct chemokinetic and chemotactic responses in neutrophil granulocytes. *European Journal of Immunology,* **8,** 1–7.

Keller, H. U., Wissler, J. H., Hess, M. W. & Cottier, H. (1977*c*). Chemokinesis and chemotaxis of phagocytes. In *Movement, Metabolism and Bactericidal Mechanisms of Phagocytes, First European Conference on Phagocytic Leucocytes,* ed. F. Rossi, P. Patriarca & D. Romeo, pp. 15–20. Padova: Piccin Medical Books.

Keller, H. U., Wissler, J. H. & Ploem, J. (1979*c*). Chemotaxis is not a special case of haptotaxis. *Experientia,* **35,** 1669–70.

König, W., Frickhofen, N. & Tesch, H. (1978). Generation and secretion of eosinophilotactic activity from human polymorphonuclear neutrophils by various mechanisms of cell activation. *Immunology,* **36,** 733–42.

Kreutzer, D. L., O'Flaherty, J. T., Orr, W., Showell, H. J., Ward, P. A. & Becker, E. L. (1978). Quantitative comparisons of various biological responses of neutrophils to different active and inactive chemotactic factors. *Immunopharmacology,* **1,** 39–47.

Lackie, J. M. & Smith, R. P. C. (1980). Interactions of leukocytes and endothelium. In *Cell Adhesion and Motility* (*Third Symposium of the British Society for Cell Biology*), ed. A. S. G. Curtis & J. D. Pitts, pp. 235–272. Cambridge: Cambridge University Press.

Lewis, W. H. (1934). On the locomotion of the polymorphonuclear neutrophiles of the rat in autoplasma cultures. *Bulletin of Johns Hopkins Hospital Baltimore,* **55,** 273–9.

Lichtman, M. A. (1970). Cellular deformability during maturation of the myeloblast. Possible role in marrow egress. *The New England Journal of Medicine,* **283,** 943–8.

Lorenz, K. (1978). *Vergleichende Verhaltensforschung. Grundlagen der Ethologie,* pp. 189–93. Wien-New York: Springer-Verlag.

McCutcheon, M. (1946). Chemotaxis in leucocytes. *Physiological Reviews,* **26,** 319–36.

McCutcheon, M., Wartman, W. B. & Dixon, H. M. (1934). Chemotropism of leukocytes in vitro. Attraction by dried leukocytes, paraffin, glass and staphylococcus albus. *Archives of Pathology,* **17,** 607–14.

Marasco, W. A., Becker, E. L. & Oliver, J. M. (1980). The ionic basis of chemotaxis. *American Journal of Pathology,* **98,** 749–68.

Miles, A. A., Miles, E. M. & Burke, J. (1957). The value and duration of defense reactions of the skin to the primary lodgement of bacteria. *British Journal of Experimental Pathology,* **38,** 79–96.

Miller, M. E. (1975). Pathology of chemotaxis and random mobility. *Seminars in Hematology,* **12,** 59–92.

Moudgil, G. C., Allan, R. B., Russell, R. J. & Wilkinson, P. C. (1977). Inhibition, by anaesthetic agents, of human leucocyte locomotion towards chemical attractants. *British Journal of Anaesthesia,* **49,** 97–105.

Naccache, P. H., Showell, H. J., Becker, E. L. & Sha'afi, R. I. (1977). Changes in ionic movements across rabbit polymorphonuclear leukocyte membranes during lysosomal enzyme release. *Journal of Cell Biology,* **75,** 635–49.

O'Flaherty, J. T., Kreutzer, D. L. & Ward, P. A. (1978). Chemotactic factor influences on the aggregation, swelling, and foreign surface adhesiveness of human leucocytes. *American Journal of Pathology,* **90,** 537–50.

Pearson, J. D., Carleton, J. S., Beesley, J. E., Hutchings, A. & Gordon, J. L. (1979). Granulocyte adhesion to endothelium in culture. *Journal of Cell Science,* **38,** 225–35.

Pfeffer, W. (1884). Locomotorische Richtungsbewegungen durch chemische Reize. In *Untersuchungen aus dem Botanischen Institut zu Tübingen,* **1,** pp. 363–482.

Pfeffer, W. (1904). *Pflanzenphysiologie. Ein Handbuch der Lehre vom Stoffwechsel in der Pflanze. Zweiter Band: Kraftwechsel,* pp. 748–826. Leipzig: Verlag von Wilhelm Engelmann.

Quie, P. G. & Cates, K. L. (1978). Clinical manifestations of disorders of neutrophil chemotaxis. In *Leukocyte Chemotaxis,* ed. J. I. Gallin & P. G. Quie, pp. 307–28. New York: Raven Press.

Ramsey, W. S. (1972*a*). Analysis of individual leucocyte behaviour during chemotaxis. *Experimental Cell Research,* **70,** 129–39.

Ramsey, W. S. (1972*b*). Locomotion of human polymorphonuclear leucocytes. *Experimental Cell Research,* **72,** 489–501.

Ramsey, W. S. (1974). Retraction fibres and leucocyte chemotaxis. *Experimental Cell Research,* **86,** 184–7.

Ramsey, W. S. & Harris, A. (1972). Leucocyte locomotion and its inhibition by antimitotic drugs. *Experimental Cell Research,* **82,** 262–70.

Rothert, W. (1901). Beobachtungen und Betrachtungen über taktische Reizerscheinungen. *Flora, oder allgemeine botanische zeitung,* **88,** 371–421.

Russell, R. J., Wilkinson, P. C., Sless, F. & Parrott, D. M. V. (1975). Chemotaxis of lymphoblasts. *Nature, London,* **256,** 646–8.

Ryan, G. B., Borysenko, J. Z. & Karnovsky, M. J. (1974). Factors affecting the redistribution of surface-bound concanavalin A on human polymorphonuclear leukocytes. *Journal of Cell Biology,* **62,** 351–65.

Schiffmann, E. & Gallin, I. J. (1979). Biochemistry of phagocyte chemotaxis. In *Current Topics in Cellular Regulation,* ed. B. L. Horecker, E. R. Stadtman, vol. 15, pp. 203–252. New York: Academic Press.

Showell, H. J., Freer, R. J., Zigmond, S. H., Schiffmann, E., Aswanikumar, S., Corcoran, B. & Becker, E. L. (1976). The structure-activity relations of synthetic peptides as chemotactic factors and inducers of lysosomal enzyme secretion for neutrophils. *Journal of Experimental Medicine,* **143,** 1154–69.

Simchowitz, L. & Spilberg, I. (1979). Chemotactic factor-induced generation of superoxide radicals by human neutrophils: Evidence for the role of sodium. *Journal of Immunology,* **123,** 2428–35.

Smith, R. P. C., Lackie, J. M. & Wilkinson, P. C. (1979). The effects of chemotactic factors on the adhesiveness of rabbit neutrophil granulocytes. *Experimental Cell Research,* **122,** 169–77.

Snyderman, R., Shin, H. S. & Hausman, M. (1971). A chemotactic factor for mononuclear leukocytes. *Proceedings of the Society for Experimental Biology and Medicine,* **138,** 387–90.

Spilberg, I., Mandell, B., Mehta, J., Sullivan, T. & Simchowitz, L. (1978). Dissociation of the neutrophil functions of exocytosis and chemotaxis. *Journal of Laboratory and Clinical Medicine,* **92,** 297–310.

Stossel, T. P. (1978). The mechanism of leukocyte locomotion. In *Leukocyte Chemotaxis,* ed. J. I. Gallin & P. G. Quie, pp. 143–60. New York: Raven Press.

Stossel, T. P. & Hartwig, J. H. (1976). Phagocytosis and the contractile proteins of pulmonary macrophages. In *Cell Motility. Book B, Actin, Myosin and Associated Proteins,* ed. R. Goldman, T. Pollard & J. Rosenbaum, pp. 529–44. Cold Spring Harbor Laboratory.

Till, G., Kownatzki, E., Seitz, M. & Gemsa, D. (1979). Chemokinetic and chemotactic activity of various prostaglandins for neutrophil granulocytes. *Clinical Immunology and Immunopathology,* **12,** 111–8.

Tinbergen, N. & Lorenz, K. (1938). Taxis and Instinkthandlung in der Eirollbewegung der Graugans. *Zeitschrift für Tierpsychologie,* **2,** 1–29.

Weissmann, G., Smolen, J. E. & Korchak, H. M. (1980). Release of inflammatory mediators from stimulated neutrophils. *New England Journal of Medicine,* **303,** 27–34.

Weller, P. F. & Goetzl, E. J. (1979). The regulatory and effector roles of eosinophils. In *Advances in Immunology,* ed. H. G. Kunkel & F. J. Dixon, vol. **27,** pp. 339–71. New York: Academic Press.

Wilkinson, P. C. (1975). Leucocyte locomotion and chemotaxis. *Experimental Cell Research,* **93,** 420–6.

Wilkinson, P. C. (1979). Synthetic peptide chemotactic factors for neutrophils; the range of active peptides, their efficacy and inhibitory activity, and susceptibility of the cellular response to enzymes and bacterial toxins. *Immunology,* **36,** 579–88.

Wilkinson, P. C. & Allan, R. B. (1978). Assay systems for measuring leukocyte locomotion: an overview. In *Leukocyte Chemotaxis,* ed. J. I. Gallin & P. G. Quie, pp. 1–24. New York: Raven Press.

Wissler, H. J., Sorkin, E. & Stecher, V. J. (1974). Regulation of serum-derived chemotactic activity by the leucotactic binary peptide system. In *Antibiotics and Chemotherapy,* ed. E. Sorkin, vol. **19,** pp. 442–63. Basel: Karger.

Woodin, A. M. & Harris, A. (1973). The inhibition of locomotion of the polymorphonuclear leucocyte by organophosphorus compounds. *Experimental Cell Research,* **77,** 41–6.

Zigmond, S. H. (1977). Ability of polymorphonuclear leukocytes to orient in gradients of chemotactic factors. *Journal of Cell Biology,* **75,** 606–16.

Zigmond, S. H. (1978). A model for understanding millipore filter assay systems. In *Leukocyte Chemotaxis,* ed. J. I. Gallin & P. G. Quie, pp. 87–96. New York: Raven Press.

Zigmond, S. H. & Hirsch, J. G. (1972). Effects of cytochalasin B on polymorphonuclear leucocyte locomotion, phagocytosis and glycolysis. *Experimental Cell Research,* **73,** 383–93.

Zigmond, S. H. & Hirsch, J. G. (1973). Leukocyte locomotion and chemotaxis. New methods for evaluation, and demonstration of a cell-derived chemotactic factor. *Journal of Experimental Medicine,* **137,** 387–410.

PETER C. WILKINSON

Peptide and protein chemotactic factors and their recognition by neutrophil leucocytes

Introduction

In this review, I wish to discuss briefly the nature of molecules which are chemotactic for leucocytes, and the selectivity of the cellular response to them. Leucocytes, which are present in large numbers in most inflammations, migrate out of the bloodstream towards foci of invading microorganisms as well as into sites of tissue damage, where they remove debris and dead cells. During the past two decades, a large number of factors chemotactic for neutrophils have been identified, including well characterized factors such as the proteins and peptides to be discussed here, or lipids derived from arachidonic acid, and also many poorly characterized chemotactic factors such as lymphokines or products released from damaged tissue cells or from microorganisms.

How much can we say at present about the recognition mechanisms used by leucocytes in their chemotactic reactions? Though many chemotactic factors identified in inflammatory exudates, cellular supernatants and other complex biological mixtures are still undefined biochemically, the structure of a number of such factors is known, and they are available in pure form, making it possible to investigate cell-surface receptors and transduction systems. One group of chemotactic factors, the formyl peptides, has been used extensively in this work (see Zigmond, this volume) and has given much valuable information. Progress is being made in defining receptors for the major complement-derived chemotactic peptide, C5a. However, chemotactic recognition cannot be explained solely on the basis of receptors for C5a or for formyl peptides. The contribution which phagocytic leucocytes make to the maintenance of the integrity of the *milieu interieur* would seem to require a capacity to monitor the environment for a wide variety of foreign or structurally disordered molecules. The recognition mechanisms involved are still badly understood by comparison with specific immune recognition. *In vitro,* phagocytes recognize, move towards, and ingest a remarkably diverse collection of molecules and particles; partly with the help of immunologically triggered activators such as complement, but frequently without them. Such diverse articles as damaged red cells, plastic beads and non-pathogenic bacteria of many species are recognized and taken up directly in the absence of antibody or complement. This versatility might, on the one hand, imply the presence of a myriad of specific recognition units on the phagocyte surface. On the other hand, selection of molecules to move towards and objects to ingest might be achieved by recognition of distinctions of a general nature which differentiate those materials from the normal tissues and fluids of the body in which the phagocyte functions. I shall review briefly the interactions of chemotactic molecules with leucocyte surface-binding sites, taking into account the peptide factors (formyl methionyl peptides and C5a) which have been much studied of late, and also protein chemotactic factors, which have been of special interest to me and which may suggest a mechanism for a 'general' form of recognition. I also wish to discuss briefly the ability of leucocytes to use specific antigen recognition by antibody as a chemotactic signal.

Leucocytes have two major functions in maintaining the *status quo* in the tissues of the body. The first is the removal of foreign microorganisms which have invaded the body, and this is usually accompanied by gross inflammation. The second is clearance of normal tissue constituents which have become damaged or aged and therefore need to be removed and replaced. If tissue damage is gross, as in the area of necrosis following a cardiac or renal infarction, an inflammatory reaction is seen. However, clearance of small areas of cell damage probably takes place continuously under physiological circumstances without an inflammatory reaction. An *in vitro* analogue of this might be the rapid removal of individual damaged red cells by nearby neutrophils in slide and coverslip preparations, the so-called necrotaxis of Bessis (1974). Similarly, non-pathogenic bacteria which gain access to the body may be cleared rapidly by phagocytic cells without overt inflammation.

In removal of pathogenic bacteria, it is likely that complement activation is often the crucial step leading to phagocytic infiltration. C5a, the major complement-derived chemotactic peptide, which attracts both granulocytes and monocytes *in vitro,* is generated at the surface of bacteria or by products released from tissue cells by either classical or alternative pathway activation. To take an example, the common pathogenic bacterium, *Staphylococcus aureus,* which produces little in the way of directly acting chemotactic factors, generates strong chemotactic activity in the presence of normal serum (Russell *et al.*, 1976). Other bacteria such as *Escherichia coli* not only possess powerful complement-activating lipopolysaccharides, but also release directly active chemotactic factors (Schiffmann *et al.*, 1974), which may be similar to the synthetic *N*-formylated chemotactic peptides. Other important factors may be products of the tissue cells themselves. In inflammatory reactions, derivatives of arachidonic acid, a normal component of eukaryotic cell membranes, may be generated by lipoxygenase and cycloxygenase activity, and some of these, such as hydroxyeicosatetraenoic acids (HETEs) or leukotriene B_4, have been documented as attractants for leucocytes (Turner, Tainer & Lynn, 1975; Goetzl *et al.*, 1980; Ford-Hutchinson *et al.*, 1980). These may prove to be important though their status remains to be evaluated. Too little is known about how they interact with cell surfaces for them to be considered in this review.

Complement activation is an efficient way of rapidly attracting many leucocytes to a focus of inflammation. When they arrive, neutrophils release a formidable array of hydrolytic enzymes and oxygen metabolites, which are themselves tissue-damaging. This may be disadvantageous, especially when the inflammation is essentially autodestructive as in immune-complex-mediated tissue damage, e.g. glomerulonephritis. Probably complement activation is also inappropriate for the physiological clearance functions of leucocytes, in which overt inflammation is undesirable. In a small focus of tissue injury, e.g. a single damaged cell or focus of denatured protein, a minimal entry of leucocytes is probably all that is required, and this may be more safely achieved by direct recognition of disorder in the damaged structure than by activation of a self-amplifying and potentially damaging enzyme cascade such as complement.

Though a number of leucocyte types i.e. neutrophils, eosinophils, mononuclear phagocytes (monocytes and macrophages) and lymphocytes show chemotactic reactions and may recognize some or all of the factors to be discussed here, the experiments to be mentioned will be, for the most part, confined to neutrophil leucocytes. These have been the easiest to work with as they are easily obtained in good purity and homogeneity, and their chemotactic, locomotor and secretory responses are rapid, vigorous and readily measured.

Formylated peptides

There have now been several reports of binding studies using a number of chemotactic ligands, and a summary of these is given in Table 1. The formylated peptides, being readily available in high purity, have been the most widely used, and aspects of their binding to neutrophils have been discussed by Zigmond (this volume). Their affinity (K_A) for neutrophil surfaces is around 10^9 l/mol (Aswanikumar *et al.*, 1977; Williams, Snyderman, Pike & Lefkowitz, 1977; Niedel, Wilkinson & Cuatrecasas, 1979), and they induce neutrophil orientation (Zigmond, 1977), chemotaxis (Showell *et al.*, 1976), exocytosis and enzyme secretion (Showell *et al.*, 1976), changes in adhesive properties (Smith, Lackie & Wilkinson, 1979), transmembrane ion fluxes (Petroski, Naccache, Becker & Sha'afi, 1979), changes in membrane potential and release of membrane-bound Ca^{++} (Gallin & Gallin, 1977; Naccache *et al.*, 1979), and metabolic activation measured by hexose monophosphate shunt activity, superoxide and hydrogen peroxide production, chemiluminescence and a variety of other activities (Hatch, Gardner & Menzel, 1978; Lehmayer, Snyderman & Johnston, 1979; Simchowitz & Spilberg, 1979; Becker, Sigman & Oliver, 1979). Thus to call them chemotactic factors is really an understatement. There are also receptor sites for formyl methionyl peptides on guinea-pig peritoneal macrophages (Snyderman & Fudman, 1980; K_A, *ca.* 10^8 l/mol; number of sites per cell, *ca.* 10^4). The type peptide is f-Met-Leu-Phe, though many similar peptides have activity and f-Nle-Leu-Phe is often used in binding studies. Important structural requirements appear to be acylation of the *N*-terminal amino group (formyl giving more activity than larger acyl groups) and hydrophobic properties, particularly in the *C*-terminal residues (Showell *et al.*, 1976). Table 2 shows data on a range of peptides studied in our laboratory (Wilkinson, 1979). The formyl methionyl peptides were most active but peptides with a variety of other *N*-terminal formylated residues also possessed activity. These peptides may be regarded as partial agonists. In addition two competitive antagonist peptides have been described. The more weakly acting is CBZ-Phe-Met (O'Flaherty *et al.*, 1978), the stronger, Boc-Phe-Leu-Phe-Leu-Phe (Aswanikumar *et al.*, 1978).

The formyl-peptide receptor, as would be expected, seems to be hydrophobic in nature. Binding is not blocked to any degree using glycosidases or proteases (Wilkinson, 1979), but is inhibited using cholesterol-specific bacterial toxins such as perfringolysin (Wilkinson & Allan, 1978*a*) and is enhanced by phospholipase A_2 (Hirata *et al.*, 1979). Binding is also increased in the presence of propanol or butanol (Liao & Freer, 1980) suggesting that cryptic receptors may be revealed upon membrane uptake of alcohols with substantial alkyl groups. It

Table 1. *Affinities of various chemotactic factors for neutrophils and the number of cell-surface binding sites*

Factor	Species of cell	K_A (l/mol) (approx.)	No. of sites	Reference
f-Nle-Leu-Phe	Rabbit	10^9	10^5	Aswanikumar *et al.*, 1977
f-Met-Leu-Phe	Human	10^8	2×10^3	Williams *et al.*, 1977
f-Nle-Leu-Phe-Nle-Tyr-Lys	Human	10^9	1.2×10^5	Niedel *et al.*, 1979
C5a	Human	2×10^8	$1\text{-}3 \times 10^5$	Chenoweth & Hugli, 1978
α_s-casein	Rabbit	2×10^5	5×10^6	Wilkinson & Allan, 1978*a*
Alkali-denatured-HSA	Rabbit	10^6	10^6	Wilkinson & Allan, 1978*a*
Native HSA [a]	Rabbit	Non-saturable low-affinity binding		Wilkinson & Allan, 1978*a*

[a] Native HSA is chemokinetic but has no chemotactic activity. The rest are chemotactic factors.

Table 2. *Comparison of chemotactic activities of peptides measured by maximally effective concentration and efficacy at that concentration* (From Wilkinson, 1979)

Peptides	No. of dose–response experiments	Maximally effective molar concentration (median value)	Efficacy expressed as chemotactic ratio (peptide/control) mean ± S.E.M.[a]
Tripeptides and larger peptides			
f-Met-Leu-Phe	8	10^{-8}	1.85 ± 0.12
f-Met-Met-Phe	5	10^{-9}	1.63 ± 0.14
f-tri-Phe	9	10^{-9}	1.46 ± 0.05
f-tetra-Phe	3	10^{-9}	1.69 ± 0.21
f-penta-Phe	3	10^{-9}	1.70 ± 0.13
f-tri-Leu	3	10^{-8}	1.40 ± 0.11
f-tri-Ala	3	10^{-9}	1.23 ± 0.06
Ac-tri-Ala	4	10^{-9}	1.48 ± 0.11
f-tri-Tyr	13	10^{-9}	1.45 ± 0.07
f-tri-Tyr Me ester	3	10^{-8}	1.32 ± 0.09
f-tri-Ser	3	10^{-6}	1.39 ± 0.01
Glu-Gly-Phe	3	10^{-4}	1.43 ± 0.13
f-Glu-Gly-Phe	2	10^{-7}	1.27 1.33[a]
Dipeptides			
f-Met-Met	11	10^{-5}	1.82 ± 0.16
f-Met-Phe	19	10^{-5}	1.52 ± 0.07
f-Met-Leu	9	10^{-5}	1.44 ± 0.07
butyryl-Met-Met	3	10^{-7}	1.46 ± 0.16
f-Tyr-Phe	4	10^{-7}	1.60 ± 0.15
f-Phe-Tyr	3	10^{-5}	1.50 ± 0.3
f-L-Leu-L-Leu	3	10^{-6}	1.38 ± 0.02
f-D-Leu-D-Leu	2	10^{-6}	1.50, 1.21[a]
f-Leu-Tyr	3	10^{-7}	1.58 ± 0.04
f-Trp-Phe	2	10^{-6}	1.12, 1.23[a]
f-Trp-Trp	2	10^{-6}	1.12, 1.20[a]
f-Tyr-Tyr	3	10^{-6}	1.59 ± 0.16
f-Ser-Met	3	10^{-6}	1.43 ± 0.07
Protein controls			
Casein (800 μg/ml)	25	—	2.36 ± 0.09
Alkali-denatured HSA (1 mg/ml = 1.4×10^{-5} M)	7	—	2.13 ± 0.20

[a] Where only two dose-response experiments were done, the figure for each is given instead of a mean ± S.E.M.

has been shown that the neutrophil receptors for formyl peptides and for C5a are distinct and that there is no cross-inhibition by either ligand in either binding or functional studies (Williams *et al.*, 1977; Chenoweth & Hugli, 1978). Further aspects of this receptor system are discussed by Zigmond (this volume).

C5a

C5a is a large peptide cleaved from the parent complement protein C5 by a C5-convertase generated when complement is activated by the classical and alternative pathways. The C5a molecule (human C5a; mol. wt 9000 approx.; 74 residues) has both chemotactic activity and activity as anaphylatoxin in releasing histamine from mast cells (see review by Hugli & Muller-Eberhard (1978)]. It is generated by tryptic cleavage and the *C*-terminal residue is therefore arginine. The arginine can itself be cleaved by carboxypeptidase B, to form $C5a_{des\ Arg}$. $C5a_{des\ Arg}$ has no activity as anaphylatoxin. It is still active as a chemotactic factor, though there has been reported to be a greater requirement for serum protein (Chenoweth, Rowe & Hugli, 1979) or for a peptide cofactor (Wissler, 1972; Perez *et al.*, 1980) than with native C5a. Tryptic enzymes other than the C5-convertase of the complement cascade can also cleave C5 to yield a variety of fragments with *C*-terminal arginines (Nilsson, Mandle & McConnell-Mapes, 1975; Orr *et al.*, 1979). These share sequences with C5a but are non-identical to it, being shorter or longer. These peptides also have chemotactic activity. They may be important biologically, since tryptic proteases may be released by a variety of cells, including neutrophils themselves (Wright & Gallin, 1977). There is presumably also a relationship between C5a and the CAT peptides (Wissler, 1972) discussed by Keller (this volume), since both are generated in similar ways, but their biochemical identity has not been investigated.

Receptors for C5a on human neutrophils are saturable and of high affinity (K_A, 1 to 3×10^8) (Chenoweth & Hugli, 1978) with approximately 10^5 binding sites per cell. $C5a_{des\ Arg}$ binds with slightly lower affinity. As mentioned, the C5a receptor is distinct from the f-Met-Leu-Phe receptor. Like formyl peptides, C5a also activates multiple functions in neutrophils, including chemotaxis (Snyderman *et al.*, 1969), hydrolase release (Goldstein, Hoffstein, Gallin & Weissmann, 1973), and metabolic activation (Goetzl & Austen 1974) *inter alia*.

A cleavage product of C5a ($C5a_{1-69}$) binds avidly to the neutrophil surface receptor but does not activate cellular functions (Chenoweth & Hugli, 1980). For cellular activation, both $C5a_{1-69}$ and the carboxyterminal sequence (residues 70–74) are required, and the two must be covalently bonded. The terminal sequence does not bind to, or compete for, the C5a receptor. Chenoweth & Hugli (1980) have, therefore, suggested that C5a binds to two cell-surface domains, one being the receptor itself, to which $C5a_{1-69}$ binds, and the other being an activation domain, to which the carboxyterminal pentapeptide binds. A cellular response is obtained only when both domains are occupied.

Amphipathic and denatured proteins and peptides

Native proteins in physiological media are not chemotactic. Protein chemotactic factors may be denatured proteins, or proteins such as caseins which normally exist in unusual physiological environments and which have been taken out of those environments (the calcium concentration in casein's normal environment, milk, is ten-fold higher than in media used for *in vitro* studies). Chemotactic activity can also be conferred on proteins by conjugation of synthetic groups, especially those of hydrophobic character (Wilkinson & McKay, 1972).

Whole casein has been used for many years as a chemotactic factor and as a stimulant of leucocyte exudates *in vivo*. The active moieties are α_s-casein and β-casein (Wilkinson, 1972, 1974). The other major casein fraction, κ-casein, does not have chemotactic activity. α_s1-Casein (the major component of α_s-casein) and β-casein are both amphipathic phosphoproteins carrying a strong net negative charge. An outline of their structures is given in Table 3. In the high-calcium environment of milk, these proteins are in highly polymerized form, the polymers being stabilized by Ca^{++}. However, in tissue culture media, where the

Table 3. *Structures of amphipathic chemotactic factors*

β-casein

209 residues. Single chain, non-glycosylated protein. No disulphide bridges.

Residues 1–50 hydrophilic, 17 negatively charged including 5 out of 5 phosphoserines; 50–209 hydrophobic, high proline, probably low α-helix.

α_s1-casein

199 residues. Single chain, non-glycosylated protein. No disulphide bridges.

Residues 1–45 hydrophobic; 46–89 hydrophilic; 90–113 hydrophobic; 114–133 hydrophilic; 134–199 hydrophobic.

Region 46–89 contains more than half the total negative charge including 7 out of 8 phosphoserines.

Succinyl-melittin

```
                               S
                               |
S–Gly–Ile–Gly–Ala–Val–Leu–Lys–Val–Leu–Thr–Thr–Gly–Leu–Pro–Ala–
Leu–Ile–Ser–Trp–Ile–Lys–Arg–Lys–Arg–Gln–Gln.
                        |       |
                        S       S
```

Alkali-denatured serum albumin

Native human serum albumin is a single-chain, non-glycosylated 3-domain molecule. Each domain is looped to form a hydrophobic core stabilized by disulphide bridges (17 in all). In alkali, the disulphide bridges are broken and the domains opened out to expose hydrophobic regions. Alkali denaturation is irreversible because of S–S bond rupture. The protein increases in viscosity and surface activity.

Ca^{++} concentration is much lower, they become soluble and able to form diffusible chemotactic gradients. They are surface-active and probably have hydrophobic regions that are exposed to aqueous solvent. Suzue, Mitsushima & Inada, (1976) showed that cyanogen bromide cleavage of the acidic from the hydrophobic moiety of β-casein resulted in loss of chemotactic activity, which was regained when the two peptides were mixed.

Succinyl bee venom melittin (Table 3) is a small peptide with similar physiochemical features to the caseins. Unconjugated melittin is amphipathic and basic, highly surface-active and highly cytotoxic. Succinylation reduces the positive charge and the cytotoxicity, and, in this form, succinyl melittin acquires chemotactic activity (Wilkinson, 1977).

Both caseins and melittin are molecules which probably have exposed non-polar groups in physiological media. In globular proteins such as albumin and haemoglobin, these groups are packed into the interior of the molecule, away from the aqueous environment. Albumin and haemoglobin are not chemotactic in their native protein form but both acquire chemotactic activity on denaturation. (Native serum albumin has a chemokinetic effect, of which more below.)

Human serum albumin (HSA) is a single-chain, non-glycosylated protein of 585 residues (mol.wt 69 000). It consists of three domains each of which is looped to form a hydrophobic cleft (which binds fatty acids and other non-polar molecules), with polar residues on the outside of the loops. The looped structure is maintained by disulphide bridges, of which there are 17 in all (Brown, 1977). Denaturation with alkali breaks disulphide bridges and opens the molecule out. Conformational unfolding is revealed by an increase in viscosity (Wilkinson & McKay, 1971), and increased exposure of hydrophobic regions is evidenced by an increase in surface activity (Wilkinson, 1974) and by difference spectroscopy (Wilkinson & McKay 1974). The chemotactic activity of the alkali-denatured protein (a-d-HSA) correlates well with these physical changes.

Haemoglobin, a folded protein with about 70% α-helix, unfolds when the haem group is removed by cold acid-acetone precipitation, the α-helix content dropping to about 50%. This unfolding is fully reversible on re-addition of the haem group and the protein regains its original conformation. Haemoglobin is not chemotactic for leucocytes, but on removal of haem it becomes so. Renaturation by addition of haemin results in a loss of chemotactic activity (Wilkinson, 1973). Thus, with haemoglobin, a reversible acquisition of chemotactic activity accompanies a reversible denaturation.

Thus there are a number of proteins with unfolded structures and exposed hydrophobic regions which are chemotactic for leucocytes. Fairly high concentrations of these proteins are required for optimal chemotactic activity (*ca.* 10^{-5}M), compared with the peptides, which are maximally active at around 10^{-8}M. Binding data for denatured albumin and α_s1-casein are shown in Table 1.

In recent, unpublished studies, we have asked whether the cell-surface-binding sites for these proteins are distinct from those for formylated peptides, and whether different proteins compete with one another for cell-surface-binding sites. Table 4 shows that f-Met-Leu-Phe and its competitive antagonist Boc-Phe-Leu-Phe-Leu-Phe do not block binding of alkali-denatured (a-d)-[^{125}I] HSA to neutrophils, whereas excess unlabelled a-d-HSA does block this binding. Furthermore, as shown in Fig. 1, both α_s- and β-casein and unlabelled a-d-HSA block binding of a-d-[^{125}I] HSA. Likewise a-d-HSA, but not the peptides, blocks

Table 4. *Binding of alkali-denatured* [^{125}I] *HSA to rabbit neutrophils in the presence of unlabelled chemotactic factors*

	μg a-d-[^{125}I] HSA added per 10^6 cells				
	46	35	24	10	5
Binding in presence of	Amount bound per 10^6 cells at 0°C, %				
Untreated cells	1.3	2.6	2.1	4.0	7.1
Boc-Phe-Leu-Phe-Leu-Phe (10^{-5} M)	2.5	2.1	3.3	5.4	6.8
f-Met-Leu-Phe (10^{-6} M)	1.6	1.8	3.3	5.7	6.2
a-d-HSA (2×10^{-4} M)	0.2	0.17	0.22	0.23	0.67

10 μg HSA ≡ 1.4×10^{-7} M.

Fig. 1. Binding of a-d-[^{125}I] HSA to rabbit peritoneal exudate neutrophils, in the absence of unlabelled protein (X), or in the presence of: unlabelled α_s-casein, 10 mg/ml (open circles); β-casein, 10 mg/ml (closed circles); or a-d-HSA, 10 mg/ml (plus signs). Note reduction of binding in presence of the chemotactic proteins. A line has not been drawn through the α_s-casein points.

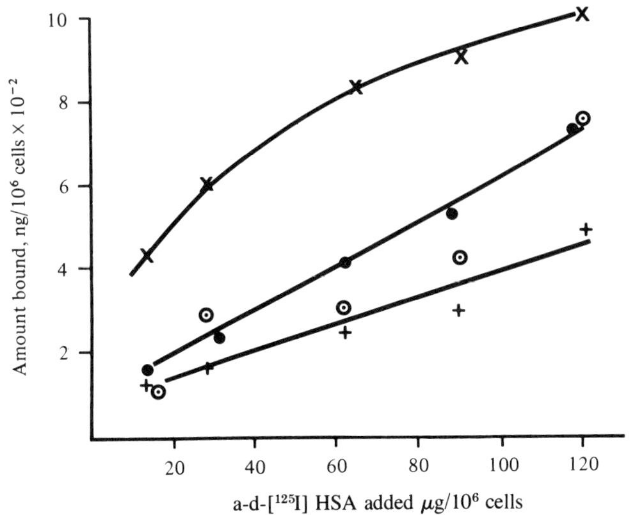

binding of β-casein to the neutrophil surface (data not shown). We conclude that the proteins, α_s- and β-casein and a-d-HSA, share binding sites which are distinct from those for the peptides.

Like the peptide receptors, the denatured-protein-binding sites are probably hydrophobic and associated with the membrane bilayer. They are protease- and glycosidase-resistant, but binding decreases on treatment of the cell with lipid-specific enzymes or toxins (Wilkinson & Allan, 1978*a*).

Chemotaxis on protein-coated surfaces

Proteins bind rapidly to artificial surfaces such as micropore filters or glass coverslips, and leucocytes crawl on protein-coated substrata. Dierich, Wilhelmi & Till (1977) showed that neutrophil leucocytes locomoting through filters would migrate as far towards casein if the casein were coated onto the filter as they would if a fluid-phase gradient of casein were set up. This suggests that leucocyte chemotaxis may be a surface-mediated phenomenon, but using the experimental conditions of Dierich *et al.* (1977), it was not possible to say whether the leucocytes were responding by chemotaxis or chemokinesis to the casein.

We set up gradients of substratum-bound proteins by allowing a gradient of fluid-phase protein to diffuse across an orientation chamber (Zigmond, 1977), then carefully washed the coverslip used to cover the chamber so that a solid-phase gradient of protein remained coated to the coverslip. We then filmed the locomotion of neutrophil leucocytes migrating on this coverslip. These leucocytes showed a significant flux towards the highest concentrations of both casein and a-d-HSA (Table 5), suggesting that leucocytes are capable of detecting and locomoting on surface-protein gradients (Wilkinson & Bradley, 1981). These studies pointed to the same conclusion as earlier work (Wilkinson & Allan, 1978*b*) using a checkerboard filter assay and various gradients of filter-bound casein in which biased locomotion towards high concentrations of surface-bound casein was also seen. These studies raise the question whether leucocyte chemotaxis is purely a surface-mediated phenomenon. This is possible, but the experiments do not provide conclusive evidence. It is possible that the protein, whether or not it is chemotactic, merely provides a suitable surface for the cell to move on and that it is irrelevant whether the chemotactic ligand is attached to the surface or not. A diagrammatic explanation of this is shown in Fig. 2.

Like the peptide chemotactic factors discussed earlier, chemotactic proteins cause release of hydrolases from neutrophils. Among these are proteases. Venge (1979) reported that casein induced release of a chymotrypsin-like cationic protease from human neutrophils and that cell locomotion was enhanced in the pre-

Table 5. *Displacement of human neutrophils on substratum-bound gradients of proteins*

	No. of cells	Mean speed (μm/min)	Mean displacement towards source (μm/min ± S.E.M.)	P	Mean displacement in axis parallel to source (μm/min ± S.E.M.)	P	Cells oriented towards gradient source in any given frame [a], %
Casein [b] gradient (original concentration 0–1 mg/ml)	25	15.3	5.3 ± 1.4	0.001	0.98 ± 1.7	NS	70
Alkali-denatured-HSA gradient (original concentration 0–8 mg/ml)	42	7.62	2.24 ± 0.51	0.0001	0.66 ± 0.62	NS	79

The displacement of the cells during periods of approximately 20 min was measured in two vectors, one perpendicular to the original gradient source, one parallel to it, by analysis of time-lapse films. Displacement parallel to the source serves as a control for displacement towards the source. With both factors, a significant flux of neutrophils towards the source was observed.

[a] Orientation was assessed by measuring the number of cells oriented in the 180° arc towards the gradient source in several arbitrarily selected frames from each film sequence.

[b] Unfractionated casein containing α_s, β, κ and other caseins. The analysis was done including only cells starting within 120 μm of the casein gradient source.

sence of this protease. It is of interest that the proteases which are released (Table 6) are able to digest chemotactic proteins such as casein and a-d-HSA readily (Wilkinson & Bradley, 1981, and Table 7). This is perhaps not surprising since these proteins, with their unfolded structures, probably provide readily accessible substrate sites. It is possible that proteolytic digestion assists the cell in locomoting on a chemotactic-protein-coated surface by freeing receptor sites for further interactions with protein.

Our conclusions about the chemotactic proteins discussed above are that several proteins with non-identical primary structures (there is little sequence homology between α_s- and β-casein, and serum albumin is unrelated to either) share the same binding sites and probably act as a group of related chemotactic factors. The leucocyte 'recognizes' the disordered structure of these amphi-

Fig. 2. (*a*). Chemotactic proteins bound to a substratum carry chemotactic sites (filled diamonds) which bind to a cell-surface-binding site. The cell migrates on the protein-coated surface and may be able to cleave the chemotactic protein and free its attachment sites by protease activity. (*b*). A cell can also move forward on chemokinetic substratum-bound protein (e.g. HSA) which has no specific cell-surface receptor. In this case, a directional response can be generated by fluid-phase chemotactic ligands, such as peptides which do not require to be bound to the substratum. If cells can move by mechanism (*b*), then locomotion is a substratum-dependent process, but chemotaxis is not necessarily so. This is supported by the observation that morphological orientation in gradients is independent of locomotion and of the nature of the substratum (Zigmond, 1977).

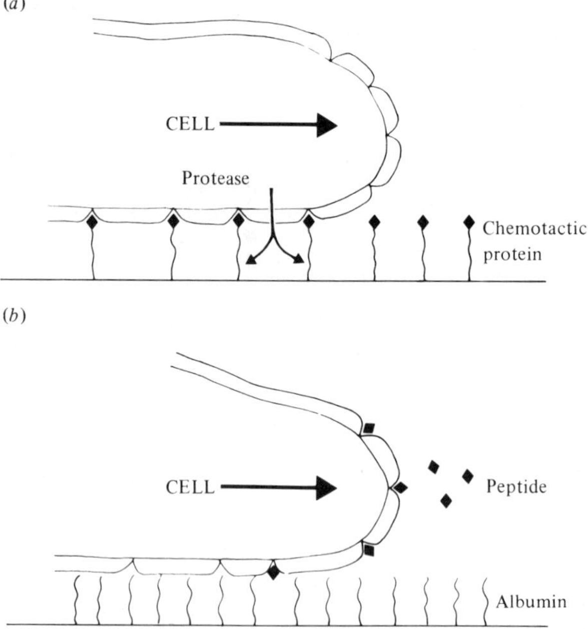

Table 6. *Protease release from human neutrophils induced by chemotactic factors*

Factor added to cells (1.2×10^6/ml)	Release as % total cellular protease content	
	Expt 1	Expt 2
None	8.4	11.0
Cytochalasin B (10 μg/ml)	8.5	—
f-Met-Leu-Phe (10^{-6}M) alone	10.0	—
f-Met-Leu-Phe (10^{-6}M) + cytochalasin B (10 μg/ml)	34.1	62.5
HSA (native) 1 mg/ml	10.3	—
HSA (native) 5 mg/ml	—	9.4
Alkali-denatured-HSA 1 mg/ml	14.7	—
Alkali-denatured-HSA 5 mg/ml	—	35.0
α_S-casein 5 mg/ml	—	30.0
β-casein 5 mg/ml	—	27.5

Protease was measured as azocaseinase activity at pH 7.2 and 37°C of the supernatant fluids after neutrophils had been incubated for 60 min with the factors at 37°C then centrifuged. The total cellular protease was measured by adding Triton X-100 (0.2%) to the cells. Note that cytochalasin B is required for protease release by f-Met-Leu-Phe but not by proteins. Note also that native HSA stimulates no protease release.

Table 7. *Protease activity in supernatant fluids of human blood neutrophils after incubation with chemotactic factors*

Cells incubated with	Protein digested per h, μg	
	Native HSA	Alkali-denatured HSA
Native HSA (5 mg/ml)	3.0	9.0
Alkali-denatured-HSA (5 mg/ml)	0.03	17.6
f-Met-Leu-Phe (10^{-6}M) + cytochalasin B (10 μg/ml)	1.4	19.0

Cells were incubated with the factors for 60 min at 37°C, then centrifuged. To the supernatants were then added either [^{125}I]HSA (native) 100 μg/ml or alkali-denatured-[^{125}I] HSA 400 μg/ml. After 18 h, 10% trichloroacetic acid (TCA) was added, the tubes were centrifuged and counts of TCA non-precipitable radioactivity were made. Note that denatured HSA is digested more rapidly than native HSA.

pathic, surface-active proteins probably without recognizing any specific, steric configuration. Though the cell possesses saturable binding sites for these proteins, this seems to be a recognition system of relatively low specificity which may be adapted to the clearance function of phagocytic cells *in vivo*.

Chemokinetic proteins

The observations on denatured HSA and other chemotactic proteins discussed above stand in contrast to observations on leucocyte interactions with the purely chemokinetic protein, native HSA (Keller, this volume). Native HSA binds to the neutrophil surface with an affinity that is too low to be determined accurately (Wilkinson & Allan, 1978*a*). It does not compete with chemotactic proteins for their binding sites, nor does it stimulate granular enzyme release (Table 6). Thus it probably does not cause transduction of a signal to the interior of the cell. We regard its chemokinetic effect as being an effect on the adhesion of leucocytes to substrata. We do not consider that the leucocyte recognizes serum albumin in its native form as such, rather that the albumin provides a substratum that allows the balanced attachment and detachment that is necessary for cell locomotion.

Antigen-specific locomotor reactions

If we now consider a highly specific recognition system rather than a system of relatively low specificity, the question may be asked whether leucocytes can use the highly selective interactions of antigen with antibody in their locomotor reactions. This is a reasonable possibility since neutrophil and eosinophil granulocytes, and mononuclear phagocytes, carry membrane receptors for the Fc portion of immunoglobulin (IgG). Thus specific antibody can bind to the leucocyte surface and can react with antigen there. Not only this, but B lymphocytes, which are motile cells, are also immunoglobulin-producing cells, and have immunoglobulin as an intrinsic membrane protein, and therefore as an intrinsic antigen receptor.

There is some evidence for antigen-specific stimulation of locomotion and chemotaxis. Jensen & Esquenazi (1975) coated neutrophils with antibody and showed that these cells, but not uncoated controls, migrated into filters in response to gradients of specific antigen. We did similar experiments using lymphocytes (Wilkinson, Parrott, Russell & Sless, 1977). Mice were primed with a protein antigen (ovalbumin, bovine serum albumin (BSA) or human serum albumin or dinitrophenylated BSA or bovine μ-globulin (BGG)]. Following challenge with the same antigen, lymphocytes were taken from the lymph node draining the site of challenge 3 to 9 days later. These lymphocytes showed positive chemotactic reactions to the priming antigen, but not to unrelated antigens, judged by filter checkerboard assays. However we were not able to determine whether it was the presence of intrinsic membrane antibody to the antigen or the presence on the cell surface of soluble antibody bound to Fc receptors, or both, which was required for the locomotor response to the antigen. These cells responded to low concentrations (in the nanogram range) of some of the antigens, e.g. ovalbumin, though the response itself was a modest one with a minority of cells moving fairly slowly into the filters. Ward, Unanue, Goralnick & Schreiner (1977) did rather similar experiments with rat lymphocytes, but in them the membrane antibody was cross-linked, not with antigen, but with anti-immunoglobulin antibody. They observed chemotactic reactions using the anti-immunoglobulin in the nanogram range, but, with higher concentrations (10 to 100 μg per ml), only chemokinetic reactions were seen (Schreiner & Unanue, 1975).

Thus there is evidence that antigen can act as a chemoattractant for antibody-coated cells. This attraction is weak and I consider it unlikely that it is effective in stimulating leucocytes to emigrate from the circulation into sites of antigen deposition. Rather, its physiological role may be to attract leucocytes, and especially lymphocytes which are already in the tissues, to antigen at short range. Macrophages are known to present antigen to lymphocytes, and this cell-cell interaction is of considerable importance in the initiation of immune responses, since macrophage-associated antigen is in many cases much more immunogenic than soluble antigen. Lymphocytes could use chemotactic locomotion to reach sources of such antigen. Clusters of lymphocytes are frequently seen around macrophages in *in vitro* cultures, though there are no good studies to suggest how they get there.

Conclusions

I think the generalization is justified that, for a leucocyte to respond directionally to a chemotactic factor, the leucocyte must possess discrete receptors for the chemotactic factor of such a type as to allow transduction of a message to the cell's interior on binding of the chemotactic ligand. The nature of the message is the subject of much current work which is outside the scope of this review. Suffice it to say that the calcium ion seems to play a key part. The 'receptor' hypothesis for chemotaxis seems reasonable since high-affinity binding allows a mechanism for gradient sensing, and for the cell to detect differences in concentration of chemotactic factors between its front and its back. It seems reasonable to suggest a structural specificity of these receptors (or some of them) for their ligands, though I have argued against this in the past. Certainly leucocytes may have evolved recognition sites for bacterial products (formyl peptides *inter alia*) or for complement. However the problem of non-specific clearance is difficult to see in these terms, and the broad-spectrum binding sites discussed above, which appear to bind several different proteins with non-identical primary structures, may be recognition sites for general features of disorder, e.g. exposed hydrophobic sites etc., rather than for any specific configurational moiety. Other quite different recognition systems based on general molecular characteristics have been postulated in other 'non-specific' immunological systems, e.g. the recognition of, and clearance of, desialated glycoproteins (Morell *et al.,* 1971). As an extension of this, red cells and microorganisms free of sialic acid are able both to activate the alternative complement pathway, and to be ingested by monocytes in the absence of complement, whereas analogous but sialic-acid-containing structuress have no activity (Fearon, 1978; Czop, Fearon & Austen, 1978). The examples discussed here suggest that leucocytes have a number of quite different chemotactic systems, since it seems at least that binding sites for formyl peptides, for C5a, for amphipathic proteins, and for antibody (the Fc receptor) are distinct. (We have unpublished evidence that the latter is distinct from the amphipathic-protein-binding site.) There may be a lot more information to come. For example, we know next to nothing about binding a chemotactic lipids or lymphokines to cell surfaces. We also do not know whether each different leucocyte type has its own sensory *Merkwelt* allowing selective chemotaxis, or whether the different distribution of inflammatory cells in different lesions is due to factors unrelated to chemotaxis.

It seems unnecessary to suggest that chemokinetic locomotion of leucocytes is necessarily receptor-mediated, though chemotactic factors in uniform concentration round the cell may stimulate chemokinesis by such a mechanism. Chemokinesis, or more accurately in leucocytes, orthokinesis, since klinokinesis has not been shown in these cells, is an operational term which covers any phenomenon that causes a change in cell speed, and is probably too imprecise to be useful at the level of molecular mechanisms. Thus metabolic stimulants or inhibitors, pH, temperature, and salt concentration, as well as factors affecting cell-substratum adhesion, all affect the speed at which cells move. There is probably no common sensor, and perhaps no sensor at all, which perceives these diverse environmental changes.

References

Aswanikumar, S., Corcoran, B., Schiffmann, E., Day, A. R., Freer, R. J., Showell, H. J., Becker, E. L. & Pert, C. B. (1977). Demonstration of a receptor on rabbit neutrophils for chemotactic peptides. *Biochemical and Biophysical Research Communications* **74,** 810–17.

Aswanikumar, S., Schiffmann, E., Corcoran, B. A., Pert, C. B., Morell, J. L. & Gross, E. (1978). Antibiotics with agonist and antagonist chemotactic activity. *Biochemical and Biophysical Research Communications,* **80,** 464–71.

Becker, E. L., Sigman, M. & Oliver, J. M. (1979). Superoxide production induced in rabbit polymorphonuclear leukocytes by synthetic chemotactic peptides and A23187. *American Journal of Pathology,* **95,** 81–97.

Bessis, M. (1974). Necrotaxis: Chemotaxis towards an injured cell. *Antibiotics and Chemotherapy,* **19,** 369–81.

Brown, J. R. (1977). Serum albumin: amino acid sequence. In *Albumin Structure, Function and Uses,* ed. V. M. Rosenoer, M. Oratz & M. A. Rothschild, pp. 27–51, Oxford: Pergamon.

Chenoweth, D. E. & Hugli. T. E. (1978). Demonstration of specific C5a receptor on intact human polymorphonuclear leukocytes. *Proceedings of the National Academy of Sciences of the U.S.A.,* **75,** 3943–7.

Chenoweth, D. E. & Hugli, T. E. (1980). Human C5a and C5a analogs as probes of the neutrophil C5a receptor. *Molecular Immunology,* **17,** 151–61.

Chenoweth, D. E., Rowe, J. G. & Hugli, T. E. (1979). A modified method for chemotaxis under agarose. *Journal of Immunological Methods,* **25,** 337–53.

Czop, J. K., Fearon, D. T. & Austen, K. F. (1978). Membrane sialic acid on target particles modulates their phagocytosis by a trypsin-sensitive mechanism on human monocytes. *Proceedings of the National Academy of Sciences of the U.S.A.,* **75,** 3831–5.

Dierich, M. P., Wilhelmi, M. P. & Till, G. (1977). Essential role of surface bound chemoattractant in leucocyte migration. *Nature, London,* **167,** 774–5.

Fearon, D. T. (1978). Regulation by membrane sialic acid of β1H-dependent decay dissociation of amplification C3 convertase of the alternative complement pathway. *Proceedings of the National Academy of Sciences of the U.S.A.,* **75,** 1971–5.

Ford-Hutchinson, A. W., Bray, M. A., Doig, M. V., Shipley, M. E. & Smith, M. J. H. (1980). Leukotriene B, a potent chemokinetic and aggregating substance released from polymorphonuclear leukocytes. *Nature, London,* **286,** 264–6.

Gallin, E. K. & Gallin, J. I. (1977). Interaction of chemotactic factors with human macrophages. Induction of transmembrane potential changes. *Journal of Cell Biology,* **75,** 277–89.

Goetzl, E. J. & Austen, K. F. (1974). Stimulation of human neutrophil leukocyte aerobic glucose metabolism by purified chemotactic factors. *Journal of Clinical Investigation,* **53,** 591–9.

Goetzl, E. J., Brash, A. R., Tauber, A. I., Oates, J. A. & Hubbard, W. C. (1980). Modulation of human neutrophil function by monohydroxyeicosatetraenoic acids. *Immunology,* **39,** 491–501.

Goldstein, L., Hoffstein, S., Gallin, J. & Weissmann, G. (1973). Mechanisms of lysosomal enzyme release from human leukocytes. Microtubule assembly and membrane fusion induced by a component of complement. *Proceedings of the National Academy of Sciences of the U.S.A.,* **70,** 2916–20.

Hatch, G. E., Gardner, D. E. & Menzel, D. B. (1978). Chemiluminescence of phagocytic cells caused by N-formylmethionyl peptides. *Journal of Experimental Medicine,* **147,** 182–95.

Hirata, F., Corcoran, B. A., Venkatasubramanian, K., Schiffmann, E. & Axelrod, J. (1979). Chemoattractants stimulate degradation of methylated phospholipids and release of arachidonic acid in rabbit leukocytes. *Proceedings of the National Academy of Sciences of the U.S.A.,* **76,** 2640–3.

Hugli, T. E. & Muller-Eberhard, H. J. (1978). Anaphylatoxins: C3a and C5a. *Advances in Immunology,* **26,** 1–53.

Jensen, J. A. & Esquenazi, V. (1975). Chemotactic stimulation by cell-surface immune reactions. *Nature, London,* **256,** 213–5.

Lehmayer, J. E., Snyderman, R. & Johnston, R. B. (1979). Stimulation of neutrophil oxidative metabolism by chemotactic peptides: Influence of calcium ion concentration and cytochalasin B and comparison with stimulation by phorbol myristate acetate. *Blood*, **54,** 35–45.

Liao, C. S. & Freer, R. J. (1980). Cryptic receptors for chemotactic peptides in rabbit neutrophils. *Biochemical and Biophysical Research Communications*, **93,** 566–71.

Morell, A. G., Gregoriadis, G., Scheinberg, I. H., Hickman, J. & Ashwell, G. (1971). The role of sialic acid in determining the survival of glycoproteins in the circulation. *Journal of Biological Chemistry,* **246,** 1461–7.

Naccache, P. H., Volpi, M., Showell, H. J., Becker, E. L. & Sha'afi, R. I. (1979). Chemotactic factor-induced release of membrane calcium in rabbit neutrophils. *Science, New York,* **203,** 461–3.

Niedel, J., Wilkinson, S. & Cuatrecasas, P. (1979). Receptor-mediated uptake and degradation of ^{125}I-chemotactic peptide by human neutrophils. *Journal of Biological Chemistry*, **254,** 10700–6.

Nilsson, U. R., Mandle, R. J. & McConnell-Mapes, J. A. (1975). Human C3 and C5. Subunit structure and modification by trypsin and C42 – C423. *Journal of Immunology,* **114,** 815–22.

O'Flaherty, J. T., Showell, H. J., Kreutzer, D. L., Ward, P. A. & Becker, E. L. (1978). Inhibition of *in vivo* and *in vitro* neutrophil responses to chemotactic factors by a competitive antagonist. *Journal of Immunology,* **120,** 1326–32.

Orr, F. W., Varani, J., Kreutzer, D. L., Senior, R. M. & Ward, P. A. (1979). Digestion of the fifth component of complement by leukocyte enzymes. *American Journal of Pathology,* **94,** 75–83.

Perez, H. D., Goldstein, I. M., Chernoff, D., Webster, R. O. & Henson, P. M. (1980). Chemotactic activity of C5a des Arg : evidence of a requirement for an anionic peptide 'helper factor' and inhibition by a cationic protein in serum from patients with systemic lupus erythrematosus. *Molecular Immunology,* **17,** 163–9.

Petroski, R. J., Naccache, P. H., Becker, E. L. & Sha'afi, R. L. (1979). Effect of chemotactic factors on calcium levels of rabbit neutrophils. *American Journal of Physiology,* **237,** C43–C49.

Russell, R. J., Wilkinson, P. C., McInroy, R. J., McKay, S., McCartney, A. C. & Arbuthnott, J. P. (1976). Effects of staphylococcal products on locomotion and chemotaxis of human blood neutrophils and monocytes. *Journal of Medical Microbiology,* **9,** 433–49.

Schiffmann, E., Showell, H. J., Corcoran, B. A., Ward, P. A., Smith, E. & Becker, E. L. (1974). The isolation and partial characterisation of neutrophil chemotactic factors from *Escherichia coli*. *Journal of Immunology,* **114,** 1831–7.

Schreiner, G. F. & Unanue, E. R. (1975). Anti-Ig triggered movements of lymphocytes: specificity and lack of evidence for directional migration. *Journal of Immunology,* **114,** 809–14.

Showell, H. J., Freer, R. J., Zigmond, S. H., Schiffmann, E., Aswanikumar, S., Corcoran, B. & Becker, E. L. (1976). The structure-activity relations of synthetic peptides as chemotactic factors and inducers of lysosomal enzyme secretion for neutrophils. *Journal of Experimental Medicine,* **143,** 1154–69.

Simchowitz, L. & Spilberg, I. (1979). Generation of superoxide radicals by human peripheral neutrophils activated by chemotactic factor. Evidence for the role of calcium. *Journal of Laboratory and Clinical Medicine,* **93,** 583–93.

Smith, R. P. C., Lackie, J. M. & Wilkinson, P. C. (1979). The effects of chemotactic factors on the adhesiveness of rabbit neutrophil granulocytes. *Experimental Cell Research,* **122,** 169–77.

Snyderman, R. & Fudman, E. J. (1980). Demonstration of a chemotactic factor receptor on macrophages. *Journal of Immunology,* **124,** 2754–7.

Snyderman, R., Shin, H. S., Phillips, J. K., Gewurz, H. & Mergenhagen, S. E. (1969). A neutrophil chemotactic factor derived from C5 upon interaction of guinea pig serum with endotoxin. *Journal of Immunology,* **103,** 413–22.

Suzue, T., Mitsushima, A. & Inada, Y. (1976). Cooperative participation of two peptides from β-casein in leukocyte chemotaxis. *FEBS Letters, Amsterdam,* **69,** 133–6.

Turner, S. R., Tainer, J. A. & Lynn, W. S. (1975). Biogenesis of chemotactic molecules by the arachidonate lipoxygenase system of platelets. *Nature, London,* **257,** 680–1.

Venge, P. (1979). Kinetic studies of cell migration in a modified Boyden chamber: dependence on cell concentration and effects of the chymotrypsin-like cationic protein of human grarulocytes. *Journal of Immunology,* **122,** 1180–4.

Ward, P. A., Unanue, E. R., Goralnick, S. J. & Schreiner, G. F. (1977). Chemotaxis of rat lymphocytes. *Journal of Immunology,* **119,** 416–21.

Wilkinson, P. C. (1972). Characterization of the chemotactic activity of casein for neutrophil leucocytes and macrophages. *Experientia,* **28,** 1051–2.

Wilkinson, P. C. (1973). Recognition of protein structure in leukocyte chemotaxis. *Nature, London,* **244,** 512–3.

Wilkinson, P. C. (1974). Surface and cell membrane activities of leucocyte chemotactic factors. *Nature, London,* **251,** 58–60.

Wilkinson, P. C. (1977). Succinyl bee venom melittin is a leukocyte chemotactic factor. *Nature, London,* **267,** 713–4.

Wilkinson, P. C. (1979). Synthetic peptide chemotactic factors for neutrophils: the range of active peptides, their efficacy and inhibitory activity, and susceptibility of the cellular response to enzymes and bacterial toxins. *Immunology,* **36,** 579–88.

Wilkinson, P. C. & Allan, R. B. (1978*a*). Binding of protein chemotactic factors to the surfaces of neutrophil leukocytes and its modification with lipidspecific bacterial toxins. *Molecular and Cellular Biochemistry,* **20,** 25–40.

Wilkinson, P. C. & Allan, R. B. (1978*b*). Chemotaxis of leucocytes towards substratum-bound protein attractants. *Experimental Cell Research,* **117,** 403–12.

Wilkinson, P. C. & Bradley, G. R. (1981). Chemotactic and enzyme-releasing activity of amphipathic proteins for neutrophils. *Immunology,* **42,** 637–48.

Wilkinson, P. C. & McKay, I. C. (1971). The chemotactic activity of native and denatured serum albumin. *International Archives of Allergy and Applied Immunology,* **41,** 237–47.

Wilkinson, P. C. & McKay, I. C. (1972). The molecular requirements for chemotactic attraction of leucocytes by proteins. Studies of proteins with synthetic side groups. *European Journal of Immunology,* **2,** 570–7.

Wilkinson, P. C. & McKay, I. C. (1974). Recognition in leucocyte chemotaxis. Studies with structurally modified proteins. *Antibiotics and Chemotherapy,* **19,** 421–41.

Wilkinson, P. C., Parrott, D. M. V., Russell, R. J. & Sless, F. (1977).

Antigen-induced locomotor responses in lymphocytes. *Journal of Experimental Medicine,* **145,** 1158–68.

Williams, L. T., Snyderman, R., Pike, M. C. & Lefkowitz, R. J. (1977). Specific receptor sites for chemotactic peptides on human polymorphonuclear leukocytes. *Proceedings of the National Academy of Sciences of the U.S.A.,* **74,** 1204–8.

Wissler, J. H. (1972). Chemistry and biology of the anaphylatoxin related serum peptide system. I. Purification, crystallization and properties of classical anaphylatoxin from rat serum. *European Journal of Immunology,* **2,** 73–83.

Wright, D. G. & Gallin, J. I. (1977). A functional differentiation of human neutrophil granules: generation of C5a by a specific (secondary) granule product and inactivation of C5a by azurophil (primary) granule products. *Journal of Immunology,* **119,** 1068–76.

Zigmond, S. H. (1977). Ability of polymorphonuclear leukocytes to orient in gradients of chemotactic factors. *Journal of Cell Biology,* **75,** 606–16.

SALLY H.ZIGMOND AND SUSAN J.SULLIVAN

Receptor modulation and its consequences for the response to chemotactic peptides

Introduction

Chemotaxis can be defined as a bias in the direction of movement of a cell or organism along the axis of a chemical gradient. The bias results from some parameter of the locomotion such as the speed of movement or the frequency, magnitude or direction of turns altering as a function of the direction of locomotion in the gradient. In order to recognize chemotaxis it is not necessary to know the mechanism. However, a knowledge of which parameter(s) is affected is necessary to understand the cell biology of the chemotactic response and frequently it is also helpful in developing reliable assays of chemotaxis. In polymorphonuclear leucocytes (PMNs), chemotaxis is due primarily to a bias in the direction of new pseudopod formation (Zigmond, 1974). Cells in a gradient are more likely to form new pseudopods toward the higher concentration of chemotactic factor than away from it. Thus, they tend to move toward the higher concentration of chemotactic factor. Any differences in the magnitude or frequency of the turns as a function of the direction of locomotion seems secondary to the directional bias of the turns. Furthermore, it appears that although the rate of locomotion is affected by the mean concentration of the chemotactic factors, it is not greatly influenced by the direction of locomotion. Consequently, neighbouring cells in a gradient move at approximately the same rate whether they are headed directly up a gradient or at a 90° deviation from the direction of the gradient. Thus, assays which measure the orientation of the locomotion of PMNs are reliable measures of the chemotactic response although a complete evaluation requires knowing the net flux or movement of cells up the gradient (see Dunn, in this volume).

In order to understand the chemotactic response, we need to know (1) how the cell detects the direction of the gradient and (2) how this information is transformed into the mechanics of directed locomotion. This paper will focus on the detection of the gradient since the molecular nature of the transduction of the information and the control of the contractile and cytoskeletal machinery inside the cell is just beginning to be worked out. For information on these aspects, the reader is referred to recent reviews by Schiffmann & Gallin (1979), and Taylor, Hellewell, Virgin & Heiple (1979).

The directional information available for chemotaxis is in the form of increasing concentrations of a chemical along the axis of the gradient. By studying the binding of chemotactic factors to their receptors we can begin to understand how the cell detects the gradient, and how the binding of chemotactic factors to their receptors results in a complicated series of events affecting a number of cellular activities.

Receptors for several different chemotactic factors have been described on PMNs. These include receptors for C5a (Chenoweth & Hugli, 1978), for a cell-derived glycoprotein (Spilberg & Mehta, 1979) for N-formylated peptides (Aswanikumar *et al.*, 1977; Williams, Snyderman, Pike & Lefkowitz, 1977) and denatured proteins (Wilkinson & Allan, 1978). Neither C5a nor the *N*-formylated peptides compete with the cell-derived factor for binding to its receptor; neither C5a nor denatured protein compete with peptide for binding to the peptide receptor. However, various *N*-formylated peptides do compete with one another and with a bacterially derived chemotactic factor for binding to the peptide receptor.

Peptide receptor

The ability of various *N*-formyl-methionyl peptides to compete with one another for binding is roughly parallel to the relative concentration required to stimulate PMN locomotion (Becker, 1979; Showell *et al.*, 1976; Wilkinson, 1979; Williams *et al.*, 1977). The concentrations at which they are effective range from 1×10^{-10} M for *N*-formyl-methionyl-leucyl-phenylalanine to 1×10^{-6} M for f-Met-Leu-Glu. The most active peptides tend to be hydrophobic but the activity depends not only on the amino acids present, but also on their order and even their stereospecificity (Showell *et al.*, 1976). If the formyl group is replaced by a hydrogen, the concentrations required to compete for binding and to cause chemotaxis increase by about four orders of magnitude (Becker, 1979). Several antagonists are known, including carbobenzoxyphenyl-alanyl-methionine (CBZ-Phe-Met) (O'Flaherty *et al.*, 1978) and t-Boc-Phe-Leu-Phe-Leu-Phe (Aswanikumar *et al.*, 1978). These reagents have no activity themselves but are able to bind to the receptor and thus compete with the *N*-formylated peptides for binding and competitively inhibit their activity.

These competition studies demonstrate that there are a limited number of receptors on a cell, i.e. the receptor is saturable. However, the actual number of receptors on a cell varies with the species. There are 5×10^4 to 1×10^5 receptors on rabbit peritoneal PMNs (Aswanikumar *et al.*, 1977; Zigmond & Sullivan, 1979). There may be slightly fewer receptors on rabbit peripheral PMNs (Becker, 1979). Human PMNs are reported to have between 2×10^3 and 2×10^4 receptors per cell (Williams *et al.*, 1977; Becker, 1979). Porcine PMNs do not seem to have peptide receptors and do not exhibit chemotaxis toward the peptides or toward bacterial extracts (Chenoweth, Lane, Rowe & Hugli 1980). Although peptide receptors do exist on human monocytes and eosinophils as well as neutrophils, they are not found on rat erythrocytes, human platelets, human circulating lymphocytes and rat brain synaptic membranes (Aswanikumar *et al.*, 1977; Williams *et al.*, 1977). There are several PMN responses mediated by binding of the chemotactic peptides. These responses can be divided into three groups on the basis of the type of information the cell appears to obtain from the receptor occupancy.

When the peptide is present in a gradient it induces oriented locomotion of PMNs, *i.e.*, chemotaxis. In contrast to bacteria, PMNs can detect the direction of the gradient without first translocating in the gradient (Zigmond, 1977). The accuracy of the orientation depends on both the concentration of the chemotactic factor and the steepness of the gradient. The orientation toward peptides is optimal in concentrations of peptide near the dissociation constant of the peptide of the receptor (Zigmond, 1977). This is consistent with the proposal that cells detect the concentration gradient by sensing *differences in the number of the receptors occupied* over some distance along the gradient (Zigmond, 1974).

Peptides present homogeneously stimulate the rate of cell locomotion, *i.e.*, they are chemokinetic. The rate of locomotion achieved depends on the peptide concentration. Usually there is a biphasic dose-response curve with the rate of locomotion increasing at low concentrations, but then reaching a concentration near the dissociation constant of the peptide receptor, where further increases decrease the rate of locomotion. At concentrations below the K_D the rate of locomotion seems to correlate with *number of receptors occupied*.

PMNs exhibit chemotaxis and locomotion over a period of hours. However, the third type of response evoked by addition of chemotactic peptides is transient. These transient responses include: increases in adhesiveness (Craddock *et al.*, 1977; O'Flaherty *et al.*, 1979), the extension of ruffles or lamellipodia (Craddock *et al.*, 1977; Zigmond & Sullivan, 1979, appendix), and release of lysosomal enzymes (Showell *et al.*, 1976). The magnitude and duration of these responses appears to depend upon the *change in receptor occupancy* that occurs upon addition of the peptides (Zigmond & Sullivan, 1979).

These responses are accompanied by a number of molecular events including changes in membrane potentials (Gallin & Gallin, 1977), ion fluxes (Naccache, Showell, Becker & Sha'afi, 1977), cyclic nucleotide levels (Hatch, Nichols & Hill, 1977), and oxidative and lipid metabolism (Goetzl & Austen, 1974; Hirata *et al.*, 1979; Pike, Kredich & Snyderman, 1979). Most of these changes occur within a few seconds after addition of peptide and terminate, or at least are greatly reduced, by 15 min. Although these responses decline the factors remain present in the medium and even after several hours continue to stimulate PMN locomotion and chemotaxis. A different response is seen if the concentration of peptide is suddenly decreased: the existing pseudopods are rapidly withdrawn and the cells bleb transiently before eventually rounding up or resuming locomotion if some stimulant remains in the medium (Zigmond & Sullivan, 1979).

Studying the interaction of the chemotactic peptide with its receptor has helped us to begin to understand these various features of the PMN response to chemotactic peptides. For example, the rapid responses upon addition of chemotactic factor suggest that the binding also occurs quickly. The rapid responses upon removing the peptide could be due to fast release of peptide from the receptor or to rapidly removing the bound receptor. Binding studies indicate that both of these mechanisms exist.

The initial rates of binding of tritiated *N*-formyl-norleucyl-leucyl-phenylalanine (f-Nle-Leu-Phe) were measured at four different concentrations of peptide. The k_{on} calculated from these initial rates according to the equation $v_i = k_{on} [R]_0[C]_0$, where v_i is the initial rate and $[R]_0$ and $[C]_0$ are the concentrations of receptor and chemotactic peptide initially present, was about 2×10^7 M^{-1} min^{-1} (Sullivan & Zigmond, 1980). The off-rate calculated from the half times of peptide dissociation as a result of washing with buffer or competition by addition of an excess of unlabelled peptide (Fig. 1) is approximately 0.6 min^{-1}.

Fig. 1. Time course of binding and dissociation of bound f-Nle-Leu-Phe at 4°C. Cells were incubated with 2×10^{-8} M f-Nle-Leu-Phe for various times and washed for 6 s to remove unbound peptide. These cells were then counted and the number of cell-associated counts were plotted as a function of incubation time (triangles). The binding of f-Nle-Leu-Phe to its receptor is rapid and reaches a steady state level within 10–15 min. Dissociation of bound peptide was assayed after they had been washed in 4°C saline for various times (filled circles) or after the addition of 2×10^{-6} M unlabelled peptide (hollow circles). Dissociation of peptide is rapid, the $t_{0.5}$ of dissociation is less than one minute.

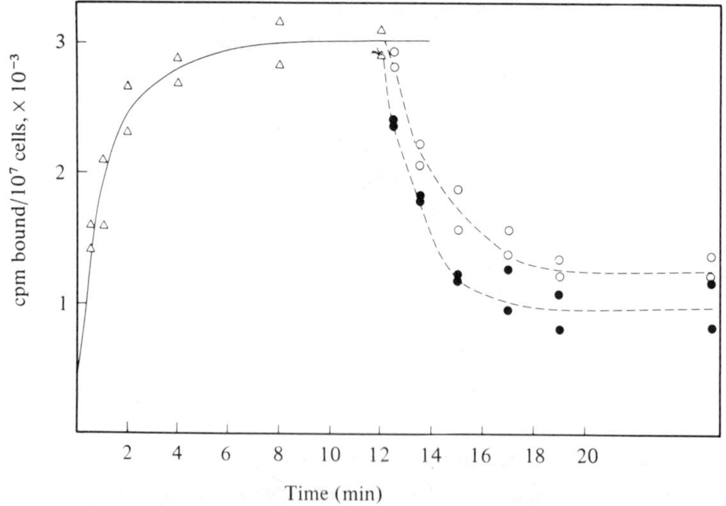

Thus the on- and off-rates of the peptide seem sufficiently rapid to account for the rapid cell responses. Further analysis of peptide binding suggests that there may be internalization and recycling of the chemotactic peptide receptor. The observations leading to this hypothesis include receptor-mediated peptide uptake, peptide-stimulated pinocytosis, receptor down-regulation and recovery which will be discussed below.

Peptide uptake

The studies described above were done at 4°C, where the binding of 1×10^{-8} M f-Nle-Leu-Phe to rabbit peritoneal PMNs reaches a plateau after about 15 min. However, at 37°C, the amount of peptide associated with the cells does not reach a plateau but continues to increase with time. Much of this cell-associated peptide is not removed by a five-minute wash which is sufficient to remove 90% of the receptor-bound peptide. Consequently, we refer to it as 'irreversibly' bound peptide. It is this 'irreversibly' bound peptide that continues to increase between 15 and 60 min while the number of reversibly bound (receptor-bound) counts remains constant (Fig. 2). The cellular location of the

Fig. 2. Time course of binding and dissociation of bound f-Nle-Leu-Phe at 37°C. Cells were incubated with 3×10^{-8} M f-Nle-Leu-Phe for various times at 37°C and then washed for 6 s. The cell-associated counts do not reach a steady state level at this temperature but continue to increase with time (filled circles). Cells were washed at 4°C with saline for various times, the counts remaining cell-associated were determined (hollow circles). The amount of peptide that is reversibly bound at 5, 15 and 30 min is approximately constant and the increase in cell-associated counts at 37°C is due to peptide which is not removed by a 5-min wash.

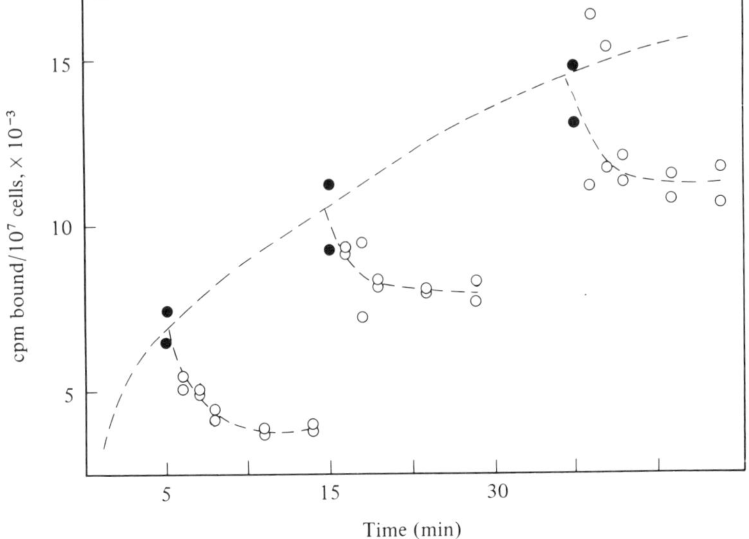

'irreversibly' bound peptide has not yet been determined. Niedel *et al.* have observed the uptake of a fluorescent peptide into pinocytic vesicles (Niedel, Kahane & Cuatrecasas, 1979). We find peptide stimulates pinocytosis by PMNs, as measured by the uptake of tritiated sucrose from the medium. In the presence of 1×10^{-7} M f-Nle-Leu-Phe sucrose uptake increases approximately four-fold from 1.7 to 6.6 nl per min per 10^7 cells.

The dose–response curve for the rate of accumulation of f-Nle-Leu-Phe at 37°C shown in Fig. 3 indicates there are two modes of peptide accumulation, one receptor-mediated and the other by bulk fluid uptake. The uptake of peptide which is saturable has a dose–response curve which reflects the binding of f-Nle-Leu-Phe to its receptor suggesting that uptake is receptor-mediated. This hypothesis is supported by the fact that CBZ-Phe-Met, a competitive inhibitor of peptide binding, also competes for saturable uptake. At concentrations below the K_D of binding of the peptide ($\sim 2 \times 10^{-8}$ M) the peptide uptake is very efficient. However, uptake continues to increase at concentrations above those which

Fig. 3. Dose–response of peptide uptake at 37°C. Cells were incubated with various concentrations of peptide at 37°C for 30 min. They were washed for 5 min at 4°C to remove the reversibly bound peptide and counts remaining cell-associated (total uptake) are plotted as a function of peptide concentration (solid circles). The amount of non-saturable uptake was measured by determining the uptake of 10^{-8} M [^{3}H]f-Nle-Leu-Phe in the presence of 1×10^{-5} M unlabelled peptide. The non-saturable uptake is proportional to the amount of peptide present and is presumably due to pinocytosis. The amount of non-saturable uptake was calculated at each concentration and then subtracted from the total uptake to give the saturable uptake (dashed line). The dose–response curve for saturable uptake closely parallels that of peptide binding to the receptor, further indicating that the saturable uptake is receptor-mediated.

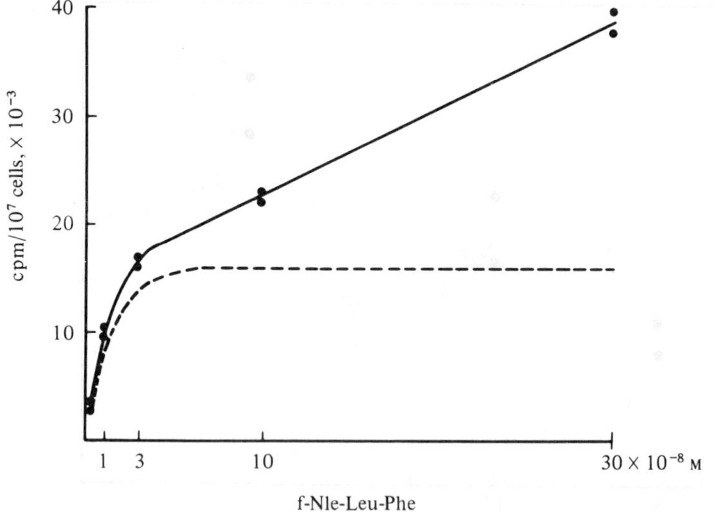

would saturate all of the receptors. This is presumably due to pinocytic uptake of the increasing f-Nle-Leu-Phe present.

Thus, at 37°C not only does the peptide bind to the receptor but it becomes associated with the cell either by binding tightly to an altered receptor and/or through compartmentalization, e.g. pinocytosis.

Modulation of receptor number

The peptide becomes cell-associated at least partially through a receptor-mediated process but the role of the receptor is unknown. It could (a) bind the peptide tightly, (b) mediate its uptake into a pinocytic vesicle, or (c) transport the peptide across the membrane. With either of the first two possibilities, the receptor would be expected to be lost along with the peptide. Indeed, cells preincubated at 37°C with unlabelled peptide and washed for 5 min show a marked decrease in their capacity to rebind tritiated peptide at 4°C. To rule out the possibility that the reduced binding was due to competition by released unlabelled peptide, the affinity of the remaining receptors was examined by Scatchard analysis. The K_D of control cells without preincubation in peptide is 2.3×10^{-8} M. The K_D of cells preincubated in 10^{-7} M f-Nle-Leu-Phe which reduced the number of binding sites by 54% is 1.9×10^{-8}. Both are within the range of error of our estimates for the K_D which is $2.2 \pm 0.3 \times 10^{-8}$ M (mean ± S.E.). If the decrease in binding capacity were due to competition by release of the unlabelled f-Nle-Leu-Phe from the cells, the affinity constant rather than the number of binding sites would be changed (Klein & Juliana, 1977).

The rate and extent of receptor loss are dependent on the concentration of f-Nle-Leu-Phe present during the preincubation (Fig. 4). Most of the receptor loss is rapid, occurring during the first five minutes, even before equilibrium binding is reached and the plateau level of receptors is achieved within 15 min.

Receptor loss is mediated by peptide binding to the specific, saturable receptor. The dose response correlates with the binding of f-Nle-Leu-Phe to its receptor with the half-maximal loss of receptors occurring at a concentration near the K_D of binding. The analogue H-Nle-Leu-Phe which does not bind to the receptor does not cause any receptor loss. Binding of the competitive inhibitor CBZ-Phe-Met also does not result in any receptor loss, but does inhibit receptor loss induced by f-Nle-Leu-Phe.

Recovery of receptor number

If cells treated with peptide and having a decreased number of receptors are subsequently washed and incubated at 37°C in medium without peptide, they recover their receptors. Cells regain 80 to 100% of their initial binding capacity within 20 min (Fig. 5). Both receptor loss and receptor recovery can occur in the absence of protein synthesis. The rate of receptor recovery upon removal of the peptide follows first-order kinetics with the available pool size being determined by the number of receptors missing from the membrane. Whether the recovery is due to the return of the same receptors or insertion of fresh receptors from a pool of preformed receptors is not clear.

Consequences of receptor loss

The loss of receptors following incubation of cells with a particular ligand has been observed in a number of endocrine systems and has been referred to as down-regulation. In some cases, down-regulation is believed to decrease the cell sensitivity. For example, the receptor loss in response to raised insulin levels accounts for some insulin-resistant states (Catt, Harwood, Aguilera & Dufau, 1979). In other cases, the consequences of receptor loss are not clear. Das & Fox (1978) have suggested that the transfer of epidermal growth factor (EGF) to the cell interior may be necessary for some of the hormone actions. In the case of the low-density lipoprotein (LDL) receptor, the receptor movements are be-

Fig. 4. Receptor loss as a function of peptide concentration. Cells were preincubated with unlabelled peptide for various times at 37°C and then washed for 5 min at 4°C to remove reversibly bound peptide. They were then incubated at 4°C with 2×10^{-8} M [^{3}H]f-Nle-Leu-Phe for 15 min to determine the number of receptors remaining. Binding capacity is plotted as a function of the preincubation time with 3×10^{-9} M (filled circles), 3×10^{-8} M (hollow circles) and 3×10^{-7} M (hollow triangles) unlabelled peptide. Control cells were preincubated without peptide (solid triangles). The receptor loss can be seen to occur rapidly. Both the rate and extent of receptor loss are dependent upon the concentration of peptide. Within 15 min the new steady state level of receptor number is reached.

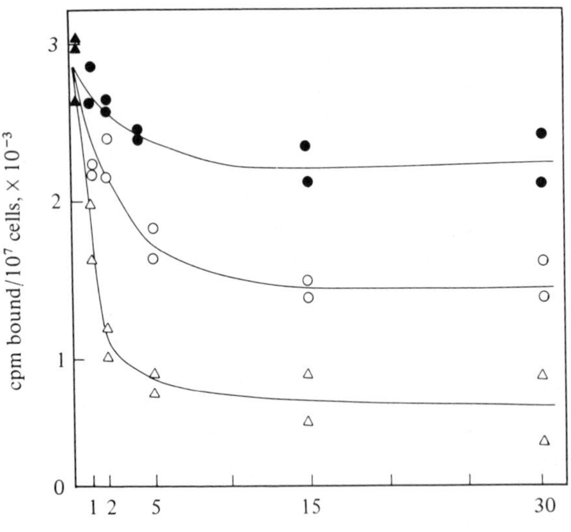

lieved to mediate transport of the LDL into the cell (Goldstein, Anderson & Brown, 1979).

The significance of the chemotactic peptide receptor loss to PMN function is not clear. The three types of responses, i.e. the transient responses induced by a change in the number of receptors occupied, the chemokinetic responses dependent upon the number of receptors occupied, and the chemotactic responses dependent on a change in receptor occupancy over one distance in the gradient, may each be affected differently. The receptor loss may have developed to limit certain transient responses such as enzyme release to situations where there is a sharp increase in concentration of the peptide. For example, addition of 2×10^{-8} M f-Nle-Leu-Phe would initially be expected to bind 50% of the total receptor or $\sim 2.5 \times 10^4$ receptors per cell. However, since this binding results in the loss of about half of the receptors the number of receptors occupied would decrease to $\sim 1.25 \times 10^4$ per cell. This could be insufficient for some of the responses. However, receptor loss does not appear sufficient to account for all of the transient

Fig. 5. Recovery of receptors after down-regulation. Cells were incubated for 20 min at 37°C with Hanks' solution (hollow triangle and circle) or 10^{-7} M unlabelled f-Nle-Leu-Phe (solid triangle and circle) and washed for 5 min at 4°C. The ability to bind 2×10^{-8} M [^{3}H] f-Nle-Leu-Phe was then tested immediately (T_0, T_{20}) or after incubation in Hanks' solution at 37°C for various times. Cells incubated with 1 μg/ml cycloheximide were subjected to the same protocol (circles). This concentration of cycloheximide decreases the rate of protein synthesis, as measured by the incorporation of [^{3}H] mixed amino acids, by 92%. Neither the loss of receptors nor their recovery appears to require protein synthesis.

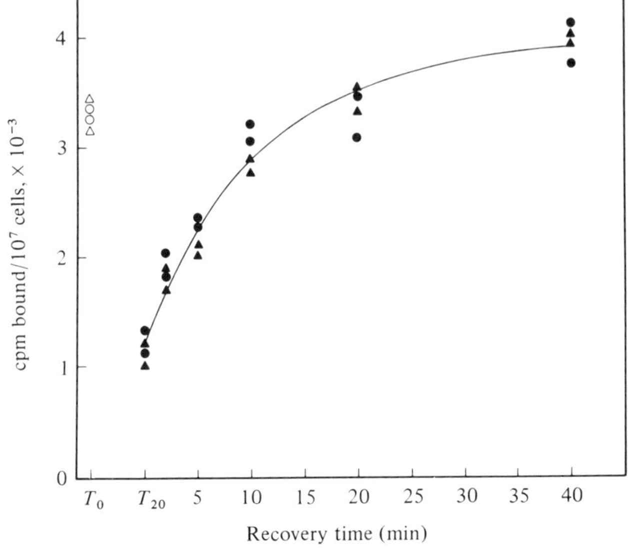

responses since 1×10^{-9} M f-Nle-Leu-Phe produces a transient aggregation of cells which is greater than that caused by 10^{-7} M after receptor loss. Yet, in 1×10^{-7} M, more receptors are occupied after receptor loss than with 1×10^{-9} M before receptor loss. Other responses, for example locomotion, may be most sensitive to low levels of receptor occupancy and thus would occur best at relatively low concentrations of peptide or after receptor loss. Indeed, immediately after addition of relatively high concentrations of peptide, the cells stop locomotion and resume it after about a 5–15-min delay when the receptor number would be expected to have decreased. After receptor loss, the maximal receptor occupancy is about one-fifth of that initially available (see Fig. 6). In fact, the rate of locomotion at different concentrations of peptide appears to correlate with the number of receptors occupied at that concentration if one corrects for the receptor loss (Fig. 6).

Cells responding to a chemotactic peptide in *in vitro* assays (and presumably also *in vivo*) are exposed to the peptide for extended periods of time, i.e. several hours. The cells in these circumstances will have reduced the number of their receptors according to the concentration of peptide present. Since orientation in a chemical gradient requires the cells to detect differences in the concentration of the chemical through differences in the number of their receptors occupied, chemotaxis should be sensitive to the number of receptors present (Spudich &

Fig. 6. Plot of number of reversible receptors occupied as a function of concentration at 4°C (solid line), or after 20 min in the various concentrations of peptide at 37°C (dashed line). The curves are drawn using the extent of receptor loss given in the text and the percent receptor occupancy from the equation $[RC] = \frac{[C]}{[C]+K_D} \times 100$, where C is the concentration of the chemotactic peptide present, RC is the fraction of receptors occupied and K_D is the dissociation constant (Zigmond, 1981).

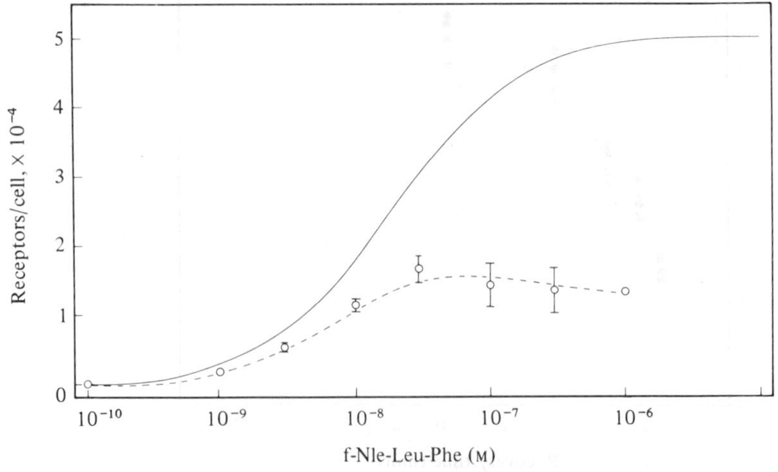

Koshland, 1975). The numbers of receptors occupied at various concentrations of f-Nle-Leu-Phe before and after receptor loss are shown in Fig. 6. These are idealized curves for peptide binding where the percent receptors occupied were calculated for each concentration, $[C]$, using the equation $\frac{[C]}{[C]+K_D} \times 100$. The K_D is 2×10^{-8} M in both cases.

Fig. 7 shows the orientation of locomotion of rabbit peritoneal exudate cells in 10-fold gradients of f-Nle-Leu-Phe over a one-millimetre bridge. The mean concentration of the peptide varied between 1×10^{-9} and 1×10^{-7} M. The peak orientation occurs at about 1×10^{-8} M and falls off more rapidly at high than at low concentrations. Also plotted is the change in the number of occupied receptors that would occur over a 10 % change in concentration in each concentration range. This should be proportional to the directional information that is available to a cell. These changes were calculated using the receptor number present after receptor loss. It is clear that under these conditions the strength of the directional information correlates well with the accuracy of the cell response.

Fig. 8 shows the calculated change in number of receptors occupied after a 10 % change in concentration before and after receptor loss. This is a measure of the magnitude of the directional signal present at the different concentrations. In both cases, the signal is maximal near the K_D and falls off at both high and low concentrations where the receptors are either mostly occupied or mostly empty. The differences between the two curves illustrates that the effect of receptor loss is to shift the optimal signal from the concentration of the K_D to a

Fig. 7. The percentage of rabbit PMNs oriented into the 180° sector toward the higher concentration of f-Nle-Leu-Phe is plotted versus the mean concentration over the one-millimetre bridge. Data, mean ± SEM; The dotted line is the change in receptor occupancy that would result from a 10% increase in peptide concentrations plotted as a function of the original concentration; this is equivalent to the magnitude of the directional signal.

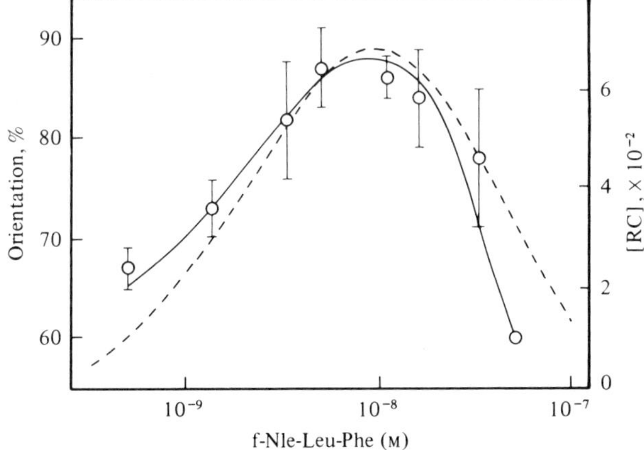

slightly lower concentration and to decrease the signal at high concentrations. This presumably accounts for the reduced orientation of cells observed at high concentrations. There is very little effect of receptor loss on the directional information available at low concentrations.

The fact that the orientation correlates with a change in number of receptors occupied over a given distance (for example 10 μm) does not define the specific details of how the cell senses a gradient. For example, the change could be detected on two different regions of the cell surface at one time or on a given region of the surface over time, e.g. on a pseudopod as it extends up or down the gradient. Nevertheless, from these data one can begin to visualize the change in receptor occupancy that occurs when 90 % of the cell population orients into the 180° toward the higher peptide concentration. For example, if we assume a regular distribution of the receptors on the cell surface we can compare the number of receptors occupied by a similar portion of the cell surface 10 μm away. Ninety per cent orientation occurs in a ten-fold gradient between 2×10^{-9} M and 2×10^{-8} M over a one-millimetre bridge (Zigmond, unpublished). The concentration at the mid-point of the bridge is 1.1×10^{-8} M. Assuming a linear gradient, the concentration 10 μm toward the higher concentration is 1.12×10^{-8} M (Zig-

Fig. 8. The calculated change in number of receptors occupied on a cell after a 10% change in concentration before (dashed line) and after (solid line) receptor loss. This is an indication of the magnitude of the directional information available to a cell in a standard gradient (e.g. a ten-fold change over one millimetre) but at different concentrations. The two curves are drawn to scale to illustrate the net effect of receptor loss. The result of receptor loss is a decrease in sensitivity at high concentrations with little effect on the directional information at low concentrations (from Zigmond, 1981).

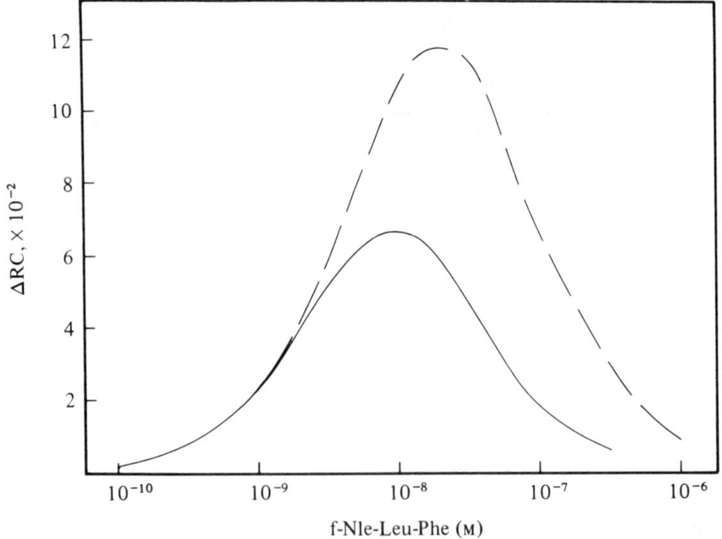

mond, 1977). At 1.1×10^{-8} M, 35.48 % of the receptors are occupied while at 1.12×10^{-8} M, 35.90 % of the receptors are occupied. We are assuming there is no change in receptor number between these two concentrations and that about 60 % of the original 5×10^4 receptors are present, i.e. 3×10^4 receptors per cell. This would be 7.5×10^3 receptors per quarter of the cell surface. This results in a change of 32 receptors (2693–2662) being occupied. At 1×10^{-9} M, where a ten-fold gradient gives about 70 % orientation, a change of only about ten receptors would be expected.

Possible receptor recycling

The receptor loss, the receptor-mediated peptide uptake, and receptor recovery after peptide removal may not be three temporally separate processes. Rather, they may all be occurring continually in the presence of peptide and result in receptor recycling. After peptide addition, receptors would be bound and internalized (or sequestered on the surface). This would result initially in the loss of receptors from the cell surface. However, within a short time there would be an accumulation of internalized receptors (or 'vacancies' in the membrane) and receptors would begin to return. When the rate of receptor loss is equalled by the rate of receptor return, the receptor number must be at its plateau level. The rate of receptor loss at this point could be equal to the rate of receptor-mediated peptide uptake which does continue to occur.

Thus, the decrease in receptor number could be merely a consequence of receptor recycling. The recycling (either of the same receptors or from a pool of receptors) could have several roles. Its function could be the removal of the peptide from the environment. Clearly, peptide is being taken up in a receptor-mediated fashion. This process would then be similar to that observed in the uptake of low density lipoprotein (LDL) by the LDL receptor (Goldstein *et al.*, 1979) and the uptake of lysosomal enzymes by the phosphomannosyl receptor (Gonzalez-Noriega, Grubb, Talkad & Sly, 1980). Perhaps more enticing in its prospects is the possibility that receptor recycling is an integral part of cell locomotion. Free receptors inserted in the front of a locomoting cell could bind peptide and induce further pseudopod extension and adhesion to the substrate. The bound receptors would then be internalized, freed of peptide and reinserted to continue to cycle. These prospects remain for further examination.

SHZ is supported by NSF grant #PCM 77-4442 and by NIH grant HL 15835 to the Pennsylvania Muscle Institute.

References

Aswanikumar, S., Corcoran, B., Schiffmann, E., Day, A. R., Freer, R. J., Showell, H. J., Becker, E. L. & Pert, C. B. (1977). Demonstration of a receptor on rabbit neutrophils for chemotactic peptides. *Biochemical and Biophysical Research Communications,* **74,** 810–7.

Aswamkumas, S., Schiffmann, E., Corcoran, B. A., Pert, C. B., Morell, J. L. & Gross, E. (1978). Antibiotics with agonist and antagonist chemotactic activity. *Biochemical and Biophysical Research Communications,* **80,** 464–71.

Becker, E. L. (1979). A multifunctional receptor on the neutrophil for synthetic chemotactic oligopeptides. *Journal of the Reticuloendothelial Society,* **26,** 701–9.

Catt, K. J., Harwood, J. P., Aguilera, G. & Dufau, M. C. (1979). Hormonal regulation of peptide receptors and target cell responses. *Nature, London,* **280,** 109–16.

Chenoweth, D. E. & Hugli, T. E. (1978). Demonstration of specific C5a receptor on intact human polymorphonuclear leukocytes. *Proceedings of the National Academy of Sciences of the U.S.A.,* **75,** 3943–7.

Chenoweth, D. E., Lane, T. A., Rowe, J. & Hugli, T. E. (1980). Quantitative comparisons of neutrophil chemotaxis in four animal species. *Clinical Immunology and Immunopathology,* **15,** 523–35.

Craddock, P. R., Hammerschmidt, D., White, J. G., Dalmasso, A. P. & Jacob, H. S. (1977). Complement (C-5a)-induced granulocyte aggregation *in vitro. Journal of Clinical Investigation,* **60,** 260–4.

Das, M., & Fox, C. F. (1978). Molecular mechanism of mitogen action: processing of receptor induced by epidermal growth factor. *Proceedings of the National Academy of Sciences of the U.S.A.,* **75,** 2644–8.

Gallin, E. K. & Gallin, J. I. (1977). Interaction of chemotactic factors with human macrophages: induction of transmembrane hyperpolarizations in macrophages. *Journal of Cellular Physiology,* **86,** 653–61.

Goetzl, E. J. & Austen, K. F. (1974). Stimulation of human neutrophil leukocyte aerobic glucose metabolism by purified chemotactic factors. *Journal of Clinical Investigation,* **53,** 591–9.

Goldstein, J. L., Anderson, R. G. W. & Brown, M. S. (1979). Coated pits, coated vesicles and receptor mediated endocytosis. *Nature, London,* **279,** 679–85.

Gonzalez-Noriega, A., Grubb, J. J., Talkad, V. & Sly, W. S. (1980). Chloroquine inhibits lysosomal enzyme pinocytosis and enhances lysosomal enzyme secretion by impairing receptor recycling. *Journal of Cell Biology,* **85,** 839–52.

Hatch, G. E., Nichols, W. K. & Hill, H. R. (1977). Cyclic nucleotide changes in human neutrophils induced by chemo-attractants and chemotactic modulators. *Journal of Immunology,* **119,** 450–6.

Hirata, F., Corcoran, B., Venkatasubramanian, K., Schiffmann, E. & Axelrod, J. (1979). Chemoattractants stimulate degradation of methylated phospholipids and release of arachidonic acid. *Proceedings of the National Academy of Sciences of the U.S.A.,* **76,** 2640–3.

Klein, C. & Juliana, M. H. (1977). cAMP-induced changes in cAMP-Binding Sites on *D. discoideum* amebae. *Cell,* **10,** 329–35.

Naccache, P. H., Showell H. J., Becker, E. L. & Sha'afi, R. I. (1977). Transport of sodium, potassium and calcium across rabbit polymorphonuclear leukocyte membrances, effect of chemotactic factor. *Journal of Cell Biology,* **73,** 428–44.

Niedel, J., Kahane, I. & Cuatrecasas, P. (1979). Receptor mediated internalization of fluorescent chemotactic peptide by human neutrophils. *Science, New York,* **205,** 1412–4.

O'Flaherty, J. T., Kreutzer, D. L., Showell, H. J., Vitkauskas, G. Becker,

E. L. & Ward, P. A. (1979). Selective neutrophil desensitization to chemotactic factors. *Journal of Cell Biology,* **80,** 564–72.

O'Flaherty, J. T., Showell, H. J., Kreutzer, D. L., Ward, P. A. & Becker, E. L. (1978). Inhibition of *in vivo* and *in vitro* neutrophil responses to chemotactic factors by a competitive antagonist. *Journal of Immunology,* **120,** 1326–32.

Pike, M. C., Kredich, N. M. & Snyderman, R. (1979). Phospholipid methylation in macrophages is inhibited by chemotactic factors. *Proceedings of the National Academy of Sciences of the U.S.A.,* **76,** 2922–6.

Schiffmann, E. & Gallin J. J. 1979. Biochemistry of phagocyte chemotaxis. *Current Topics in Cell Regulation,* **15,** 203–61.

Showell, H. J., Freer, R. J., Zigmond, S. H., Schiffmann, E., Aswanikumar, S., Corcoran, B. & Becker, E. L. (1976). The structure activity relations of synthetic peptides as chemotactic factors and inducers of lysosomal enzyme secretion for neutrophils. *Journal of Experimental Medicine,* **143,** 1154–69.

Spilberg, I. & Mehta, J. (1979). Demonstration of a specific neutrophil receptor for a cell-derived chemotactic factor. *Journal of Clinical Investigation,* **63,** 85–88.

Spudich, J. L. & Koshland, Jr, D. E. (1975). Quantitation of the sensory response in bacterial chemotaxis. *Proceedings of the National Academy of Sciences of the U.S.A.,* **72,** 710–13.

Sullivan, S. J. & Zigmond, S. H. (1980). Chemotactic peptide receptor modulation in polymorphonuclear leukocytes. *Journal of Cell Biology,* **84,** 703–11.

Taylor, D. L., Hellewell, S. B., Virgin, H. W. & Heiple, J. (1979). The solation-contraction coupling hypothesis of cell movement. In *Cell Motility: Molecules and Organization,* ed. S. Hatano, H. Ishikawa & H. Sato. Tokyo: University of Tokyo Press.

Wilkinson, P. C. (1979). Synthetic peptide chemotactic factors for neutrophils, the range of active peptides, their efficacy and inhibitory activity, and susceptibility of the cellular response to enzymes and bacterial toxins. *Immunology,* **36,** 579–88.

Wilkinson, P. C. & Allan, R. B. (1978). Binding of protein chemotactic factors to the surfaces of neutrophil leukocytes and its modification with lipid-specific bacterial toxins. *Molecular and Cellular Biochemistry,* **20,** 25–32.

Williams, L. T., Snyderman, R., Pike, M. C. & Lefkowitz, R. J. (1977). Specific receptor sites for chemotactic peptides on human polymorphonuclear leukocytes. *Proceedings of the National Academy of Sciences of the U.S.A.,* **74,** 1204–8.

Zigmond, S. H. (1974). Mechanisms of sensing chemical gradients by polymorphonuclear leukocytes. *Nature, London,* **249,** 450–2.

Zigmond, S. H. (1977). The ability of polymorphonuclear leukocytes to orient in gradients of chemotactic factors. *Journal of Cell Biology,* **75,** 606–16.

Zigmond, S. H. (1981). Consequences of chemotactic peptide receptor modulation for leukocyte orientation. *Journal of Cell Biology,* **88,** 644–47.

Zigmond, S. H. & Sullivan, S. J. (1979). Sensory adaptation of leukocytes to chemotactic peptides. *Journal of Cell Biology,* **82,** 517–27.

PETER C. NEWELL

Chemotaxis in the cellular slime moulds

Introduction

The phenomenon of chemotaxis is seen as a recurrent theme throughout all phases of the life cycle of the cellular slime moulds. During the course of gathering food the phagocytic amoebae are chemotactically attracted to folic acid produced by bacteria growing (in their native habitat) in the top layers of forest litter; negative (or repulsive) chemotaxis between the amoebae during this phase aids in dispersal of the amoebal population. In contrast when food is no longer to be found, the need for such dispersal is reversed and the amoebae use chemotaxis to swarm towards collecting centres. Under some conditions, if opposite mating types are present, pheromones are liberated and a zygotic macrocyst is formed. With populations of a single mating type, however, an impressive long-range chemotactic system is set in motion in which rhythmic pulses of a chemoattractant are produced by centres and relayed outwards throughout the amoebal population. This leads to waves of inward motion visible to the unaided eye and to streams of amoebae moving towards the centres. After a few hours the amoebae at these centres collect into a multicellular body covered by a glistening sheath that can migrate in a slug-like fashion over the substratum in search of a suitable place to start differentiation. After this migration the slug goes through a series of changes of shape and form that, from recent evidence, involve differential chemotactic movement and paralysis of cells until finally the characteristically shaped fruiting body is produced (Fig. 1).

Chemotaxis towards food

One of the most basic needs of an amoeba is to find food. Bacteria form the principal food material in the wild and a chemotactic system that detects a diffusible substance given off by bacteria would be highly advantageous. Pan, Hall & Bonner (1972, 1975) found that drops of vegetative amoebae placed on small cellophane squares on agar were strongly attracted to small drops of folic acid placed on the agar adjacent to the amoebae. The rate at which cells moved off the cellophane square was used as an index of attraction to various test substances and the results of Pan *et al.* clearly indicated that only substances possessing the pteridine ring showed activity for the species *Dictyostelium discoideum, Polysphondylium pallidum* and *P. violaceum*. Activity varied with different substitution of the pteridine ring, with folic acid active for *D. discoideum* over a wide range down to nanomol concentrations. Although vegetative

Fig. 1. The life cycle of the cellular slime mould *Dictyostelium discoideum*. The spores germinate to produce amoebae which are normally haploid and about 10 μm in diameter. These amoebae engulf soil bacteria in their natural forest environment or, under laboratory conditions, they grow on *Klebsiella aerogenes* on agar-solidified media.

The aggregation phase is initiated by starvation. A few of the population of starving amoebae start producing pulses of cAMP that cause periodic movement of the responding amoebae towards the pulsating centre. These amoebae then relay the signal outwards from the centre, thereby forming outwardly moving bands of moving and stationary amoebae. The inward radial motion of the amoebae is unstable and eventually breaks down to produce streams of amoebae that finally move into the aggregation centre. A slime sheath is produced that covers the aggregate as the aggregation phase comes to an end.

During the morphogenetic phase the aggregate, which may be composed of up to 100 000 cells, is transformed into a migrating slug by upward movement of the amoebae within the slime sheath and a bending movement of the resultant 'finger' back down on to the substratum. The slug, which is responsive to heat and light gradients, can migrate for up to 20 days (depending on environmental conditions). It eventually stops moving and transforms itself into a sombrero-like body and then by means of cell movements and elongation, into a round mass of spores elevated on a slender cellulose-ensheathed stalk. (From Newell (1978*a*).)

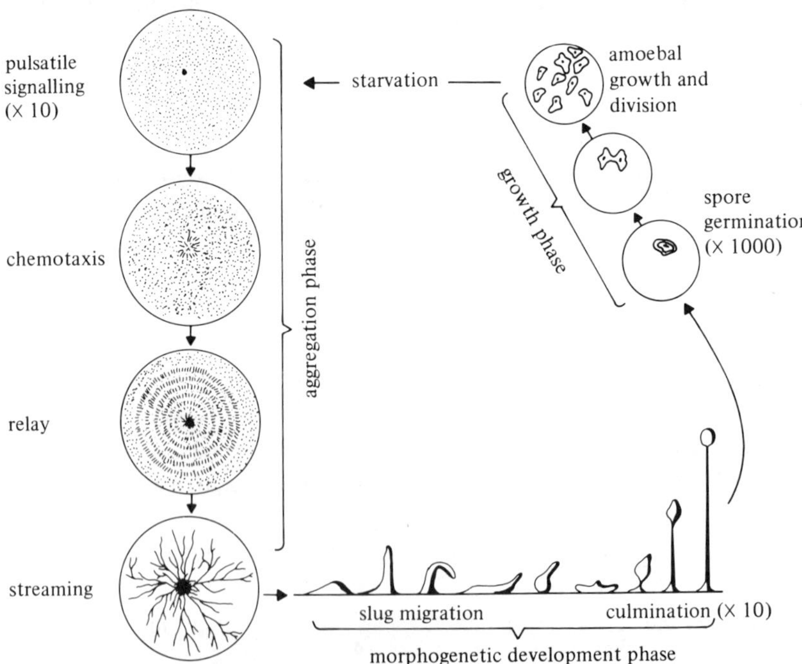

cells rapidly showed a response to folic acid, starving cells in the aggregation phase were very much less active; this argues strongly against folic acid being an aggregation-phase attractant. Moreover, although cellular slime moulds contain appreciable amounts of folic acid, they require the substance as an essential vitamin in defined liquid media (Franke & Kessin, 1977) and presumably they normally rely on bacteria for their dietary needs. This vitamin, which may be liberated by bacteria, would, therefore, be a very suitable compound with which to detect a potential food source for the amoebae.

It has been found that the amoebae of *D. discoideum* and *P. violaceum* excrete an enzyme that rapidly destroys folic acid. Its function is, presumably, to decrease the background concentration of the compound and thereby aid sensitivity of the amoebal folate receptors. Work by Pan & Wurster (1978) has indicated that the enzyme catalyses the hydrolytic deamination of folic acid to 2-deamino folic acid (Fig. 2). Curiously, for some species the deaminated product is still active and such species inactivate the folate by breaking the link between the pteridine ring and *p*-aminobenzoyl glutamate (Kakebeeke, de Wit & Konijn, 1980).

The actual mechanism of chemotaxis brought about by folic acid is not yet known in any detail, but work by Wurster, Schubiger, Wick & Gerisch (1977)

Fig. 2. The structures of folic acid and cAMP which are used as chemoattractants in the growth phase and aggregation phase (respectively) of *D. discoideum*.

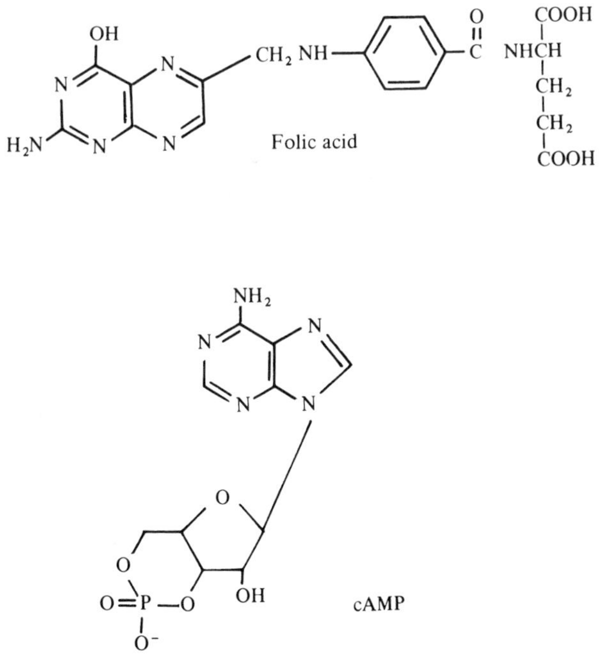

Folic acid

cAMP

and Mato *et al.* (1977*b*) has indicated that for *D. discoideum* folic acid can trigger the activity of the enzyme guanylate cyclase and can thereby produce a brief internal pulse of cyclic guanosine monophosphate (cGMP) in much the same way as external cyclic adenosine monophosphate (cAMP) signals can cause such effects in aggregating cells (see below). Indeed at early times of starvation it is reported that amoebae are sensitive to both folic acid and cAMP in this way, and it appears likely that the two receptors for these compounds share a common membrane transducing system that links them to the internal chemosensory apparatus.

Negative chemotaxis

The notion that vegetative amoebae of the cellular slime moulds repel each other is an old one, but until recently has attracted little attention. The earliest observations (reviewed by Bonner, 1977) were those of Samuel (1961) who positioned a small agar-coated square of cellophane near to a drop of vegetative amoebae of *Dictyostelium mucoroides* on agar. He found that the amoebae moved in all directions from the drop and some went on to the cellophane square. If, after a while, he turned the square through 90° in the horizontal plane so that a different edge was now closest to the expanding drop of amoebae, he noticed that the amoebae on the cellophane changed their direction through 90° and continued to move away from the original drop.

More recent results by Keating & Bonner (1977) confirmed this negative chemotactic ability. Using *D. discoideum* and *D. mucoroides* they showed that dense drops of vegetative amoebae expanded much faster if the agar on which they were deposited was only 0.5 mm deep rather than 4 mm deep. The simplest explanation was that in the shallow agar a repellent produced by the amoebae was less diluted than with the deep agar (Fig. 3(*a*)). The repellant hypothesis was also tested by placing two drops of *D. discoideum* amoebae side-by-side on the agar and observing the pattern of movement. After a few minutes much more spreading had occurred on the outside edges of the two drops than on the edges between the drops (Fig. 3(*b*)).

More recently, Kakebeeke, de Wit, Kohtz & Konijn (1979) have employed a similar system of two drops of amoebae on agar to develop a quantitative assay for the repellent. The assay is similar to the positive attractant assay of Konijn (1970) but uses a 0.1-μl drop of amoebae at low population density placed on hydrophobic agar alongside (within 100 μm of) another drop of amoebae at high population density. The hydrophobic agar inhibits amoebae moving beyond the original boundary of the drop. Kakebeeke *et al.* found that if the high-density drop contained vegetative amoebae of *Dictyostelium* or *Polysphondylium* species then within a few minutes of deposition on the agar the amoebae in the low-density drop moved to the furthest edge of the drop away from the cells at high density. The best effect was seen if amoebae of the same species were used for responding and producing populations (possibly implying different species-specific repellents) and, while starving, amoebae preparing to aggregate would act as responders but did not themselves produce detectable amounts of repellent. Homogenates of the high-density vegetative amoebae could replace the live cells and by methanol extraction of the amoebae a crude cell-free preparation of repellent was obtained. Its chemical nature has not yet been elucidated but it appears to be a small (dialysable) molecule that is inactivated by boiling at 100°C for 10 min and is sensitive to inactivation by acid at room temperature. Cell homogenates (but not boiled homogenates) also inactivate the repellent, suggesting that the compound is enzymatically degraded.

Presumably the function of such a negative chemotactic system would be to

Fig. 3. Demonstration of negative chemotaxis. Fig. 3(*a*) shows the spreading of growth-phase amoebae on agar surfaces of different depths. Note that with shallow agar the amoebae spread farther from the drop. Fig. 3(*b*) shows the effect of placing two drops of growth-phase amoebae close to each other on an agar surface. Note that spreading has occurred to a much greater extent on the outer edges than between the two drops. (Redrawn from Keating & Bonner (1977).)

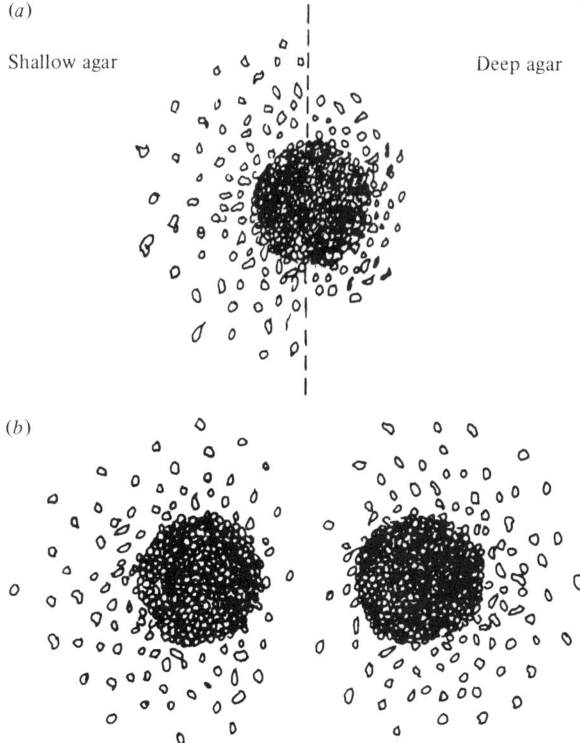

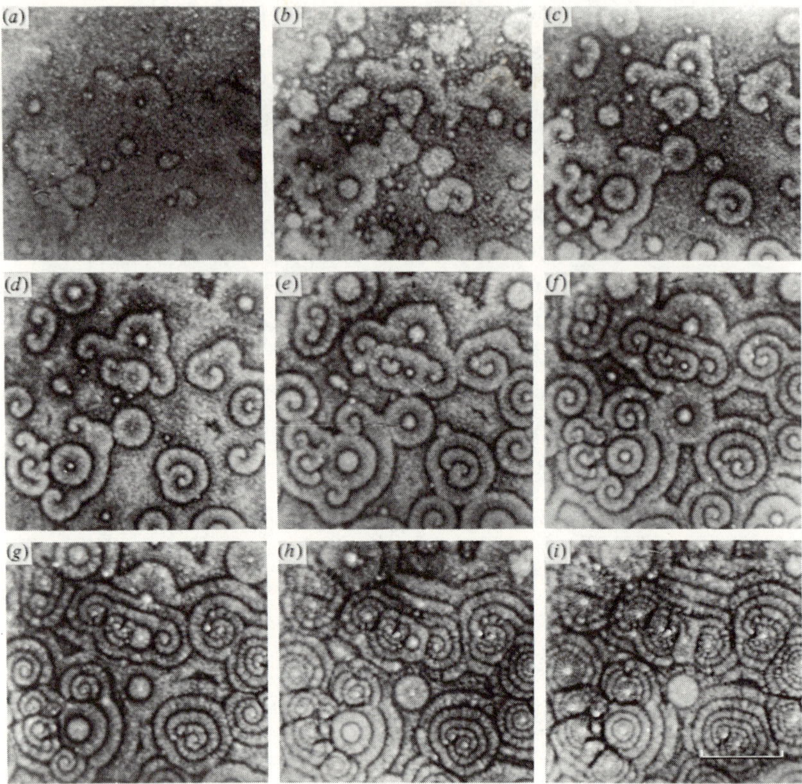

Fig. 4. Time-lapse photographs of aggregating fields of amoebae of the cellular slime mould *D. discoideum* showing spiral and concentric waves. The nine frames were taken at 10-min intervals. cAMP is produced in the form of brief pulses that emanate from the attraction centres. The signals are relayed outwards at approximately 300 μm min^{-1} at the cell density used. Besides relaying the pulses, the amoebae respond to the periodic signals by moving inwards towards the signalling centres for a short distance (at a velocity of approximately 12 μm min^{-1}) each time a pulse reaches them. This discontinuous movement has been visualized in these photographs by the use of a darkfield optical system that reveals the moving amoebae (which are elongated) as bright bands and the stationary amoebae (which are more rounded) as dark bands. The particular strain of slime mould shown (NP 368) is a streamer mutant that has wide movement bands that aid the visualization of waves. Waves can be either of a concentric form, the periodicity of which is controlled by pace-maker amoebae at the signalling centres or alternatively may take a spiral form in which periodicity is governed by the interval during which the amoebae are refractory to stimulation after receipt of a signal pulse. The scale bar represents 10 mm. Note that centres compete for territory and can be seen to swamp each other. See, for example, the concentric field at the top centre of each frame that is signalling strongly through frames (*a*)–(*g*) but is swamped by frame (*h*). The prominent concentric field that may be seen in the centre of

ensure that feeding amoebae spread out efficiently over the substratum. When the repellent has been purified it will be interesting to see whether it has a receptor system that exerts its effects via formation of cyclic nucleotides like that for cAMP and folic acid. Clearly the movement response is likely to be similar in positive and negative chemotaxis: it is the direction of movement to or from the stimulus that differs.

Chemotaxis during aggregation

The most studied chemotactic ability of *Dictyostelium* is that occurring during the aggregation phase when starving amoebae collect together and form a multicellular body. This curious pattern of behaviour is best observed on non-nutrient agar surfaces and is amenable to analysis by visual observation, by genetic manipulation and by biochemical assay.

Visual observations

For the first 4–6 h after the initiation of starvation no oriented movement is observable, but during the 4–5 h that follow, the cells begin to move towards collecting centres. Their progress is not continuous but occurs in rhythmic steps of about 100 sec duration that are separated by quiescent phases of 3–4 min in which the amoebae remain relatively still and await the next signal to move forward. In dense populations this periodic movement results in the appearance of dark and light bands under darkfield illumination (Fig. 4). Such bands, which have been studied by a number of observers over many years (Arndt, 1937; Bonner, 1944; Shaffer, 1957; Gerisch, 1968; Alcântara & Monk, 1974; Gross, Peacey & Trevan, 1976), result from the slight differences in optical properties of moving and stationary amoebae that make moving cells look bright and stationary cells look dark under carefully aligned darkfield conditions. The bands (and hence the relayed chemotactic signals) move outwards at roughly 300 μm min^{-1} (with population densities of 10^5 amoebae per cm^2) and under certain conditions fields of more than three centimetres in diameter may be observed with 15 or more bands moving outward. The width of the bright bands is determined by the duration of the movement period after the amoebae have been chemotactically stimulated. In wild-type populations this bright band width is fairly constant, the movement period being a controlled time of about 100 s

Fig. 4. (*continued*)
each frame is signalling more slowly than the surrounding spiral centres and gradually loses territory on all sides.

(This photograph by P. C. Newell, F. M. Ross and F. C. Caddick was originally published in '*From Being to Becoming*' by Ilya Prigogine, published by W. H. Freeman & Co. 1980.)

during which they can move (at a velocity of 12 μm min^{-1}) about 20 μm (or roughly two cell diameters). In some mutant strains, however, this movement phase has been greatly extended and in such strains the bright band continues until the next chemotactic signal wave abruptly stops the cells and restarts them once again (F. M. Ross & P. C. Newell, unpublished).

In normal wild-type cells the width of the dark stationary band is determined by the frequency of pulsations emanating from the collecting centre. The higher the frequency the narrower are the dark bands. The frequency is not constant throughout the course of aggregation but starts at roughly one pulse every 10 min then decreases to once every 5 min and finally to once every 2–3 min before becoming indistinct (and possibly) continuous during the later stages of aggregate formation. Such observations have led Durston (1973) to postulate a 'gating' mechanism that controls responses to a signal-propagation system or 'pacemaker' in the aggregation centre. Such a gating system could operate by means of the refractory state that cells are seen to enter after they have experienced a chemotactic pulse. Although amoebae that have received a chemotactic signal are able to respond chemotactically to another signal within a few seconds, they cannot relay the new signal for several minutes (the relay refractory period). Such a refractory period, which Durston found to be greater than 6 min before 6 h of development, decreases to less than 3 min after 9 h.

The light and dark bands seen under darkfield optics may either be concentric or spiral (Fig. 4). Detailed observations by Durston (1974) and Gross *et al.* (1976) suggest that spiral aggregation territories may arise by breakage of a concentric ring, followed by propagation of the signal from this break point as a continuous two-dimensional spiral. Once initiated, such a spiral wave no longer depends on the centre for its periodic initiation but propagates itself continuously, the apparent frequency of the wave experienced by a cell positioned in the field being simply limited by the refractory period for signal relay of the cells. Consequently, spiral wave territories generally 'beat' faster than concentric ones and commonly swamp them where centres are competing for territory (see Fig. 4).

The initial radially inward motion of the amoebae toward the collecting centres at some stage becomes unstable and streams of cells are produced rather than a uniform field (Fig. 5). The instability of the radial motion is due to the amoebae moving chemotactically to the locally produced signal. This is a signal relayed from the collecting centre but in populations of amoebae that show slight inhomogeneity of distribution the inhomogeneity is reinforced and amoebae collect into streams and leave areas devoid of cells. This tendency to form streams that

Fig. 5. Transition from concentric and spiral waves to stream formation. The optical conditions were as for Fig. 4. The time interval between frames was approximately one hour. The scale bar represents 10 mm.

(a)
(b)
(c)

often moved at right angles to (or even at times away from) the final direction of movement was one of the clues that originally suggested that aggregation territories were achieved by a relay system rather than by a large shallow gradient from the centre (Bonner, 1949; Shaffer, 1957, 1958).

Genetic analysis of chemotactic mutants

If the cellular slime moulds are grown as clones on bacterially inoculated agar plates, the most easily identified mutant phenotype is the 'aggregateless' type. Because of the failure of such mutants to aggregate, further cell interaction is blocked and the characteristic appearance of aggregates undergoing development to fruiting bodies fails to occur (Fig. 6).

The number of genes involved in the aggregation process and their position on the seven linkage groups has been investigated in *Dictyostelium* by Williams & Newell (1976) and Coukell, (1975, 1977) and in the genus *Polysphondylium* by Warren, Warren & Cox (1975, 1976). For these studies, use was made of the parasexual cycle. In this cycle, haploid amoebae occasionally fuse to form diploids which can be selected and isolated. Subsequent spontaneous haploidization of these diploids leads to segregation of chromosomes and allows identification of linkage groups (see Newell, 1978*b*, for a review of slime-mould genetics). Haploid mutants were fused under conditions of starvation and the diploids formed (at a frequency of about 2×10^{-5} in 24 h) were selected using temperature sensitivity for growth. By analysis of the aggregational abilities of the diploids isolated in this way, it was clear that of the order of 50–100 genes were essential specifically for the aggregation process. Segregation of these diploids has, so far, revealed no particular gene clustering and mutations have been found on all of the 6 well marked linkage groups.

One of the objectives of the genetic analysis is to correlate genetic mutations with lesions in known compounds of the chemotactic apparatus. One such correlation that has recently been made by Brachet and colleagues is that mutants bearing mutations in the gene *pdsA* (located on linkage group IV) fail to produce phosphodiesterase enzyme needed to destroy the cAMP signal (see below). The mutant phenotype is aggregateless but they will aggregate if exogeneous phosphodiesterase enzyme is added to the amoebae starving on agar (Brachet, Dicou

Fig. 6. Patterns of aggregation and development seen in clones of *D. discoideum* growing on lawns of bacteria. (*a*) Wild type showing an expanding ring of amoebal growth behind which the starving cells proceed through the development phase. (Waves and streams are not easily seen under these very-high-density growth conditions.) (*b*) A streamer mutant that aggregates and develops but which (compared to the wild type under these conditions) forms huge streams leading into the aggregates – an example of an 'aggregation pattern mutation'. (*c*) An aggregateless mutant that grows but does not aggregate or develop. The scale bar represents one millimetre.

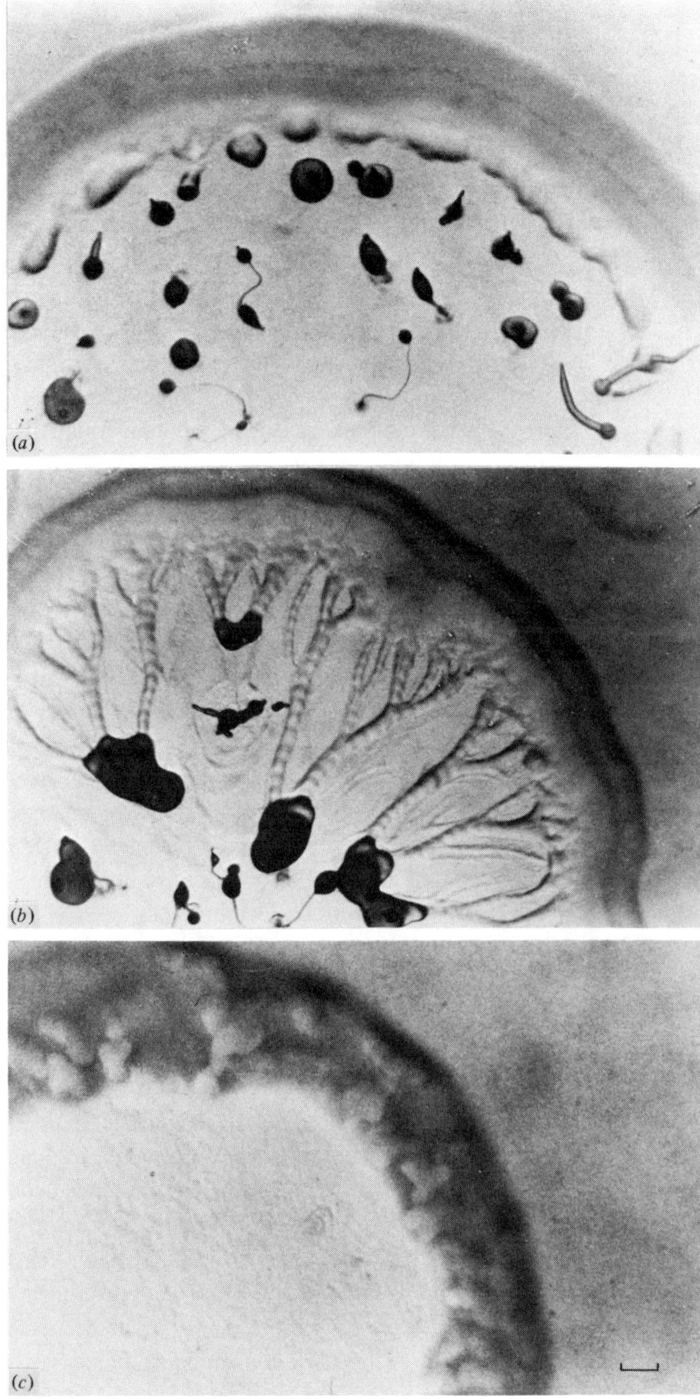
(a)
(b)
(c)

& Klein, 1979; Barra, Barrand, Blondelet & Brachet, 1980). It remains to be determined whether the mutation is in a structural gene for the enzyme or at a regulatory locus concerned with phosphodiesterase production.

Due to the sequential nature of the aggregational (and later developmental) process, the aggregateless mutants show pleiotropic defects such that all events that would normally occur after the stage that is blocked fail to occur. This presents difficulties when attempts are made to correlate genetic lesions with biochemical deficiencies. To circumvent this problem mutants have been studied that are typified by a disturbed *pattern* of aggregation but which allow development to proceed normally. Such 'aggregation pattern' mutants may show, for example, very large streams and hence large territory sizes (Fig. 6) or conversely little or no streams (without signal relay being operational) and hence very small fruiting bodies. Other examples of this group of mutants could, in principle, include those with abnormally high frequency of pulses produced by the aggregation centre and mutants with an extended relay refractory period leading to infrequent pulses of signal being relayed.

The large stream class ('streamers') have been analysed genetically by Ross & Newell (1979). Twenty-four haploid mutants were fused together in pairs and the pattern of complementation determined. The mutants were found to occur in seven distinct complementation groups located on three linkage groups. From their phenotypic behaviour, the mutants clearly possess a greater-than-normal ability to suppress further initiation of centres once a local centre has formed in a growing clone. Because of the continuous outward growth of the clone on a lawn of bacteria and the continuous production of newly starving amoebae as the growth ring expands, aggregation centres once formed tend to dominate the newly starving amoebae and entrain them into the older, dominating centre. In this way very large streams can be formed in the mutant clones. In wild-type clones this dominance is somehow regulated to allow new centres to be formed. The biochemical basis of the dominance effect is as yet incompletely understood. The mutants appear to resemble the wild type in every respect tested with the exception that they seem to persist in the movement phase for longer than the wild type after receiving a chemotactic signal (F. M. Ross & P. C. Newell, unpublished). This leads to the formation of a wide (bright) movement band seen under darkfield optics. A possible connection between such a prolonged movement response and domination of one centre by another in clones could be that cells might not be able to initiate centre formation while chemotactically responding to a signal from another centre. This would have the effect of greatly diminishing the tendency of the streamer mutants to initiate new centres in clones, compared to the wild type which has a period without chemotaxis (the wide dark band) following each chemotactic movement response.

Biochemical mechanism of chemotaxis

For the three species of *Dictyostelium, D. rosarium, D. purpureum* and *D. mucoroides,* the molecule that is rhythmically emitted by the aggregation centres is cAMP (Konijn, 1972; and Fig. 2). Other species use different attractants that are less well characterized. *P. violaceum* and *P. pallidum* amoebae from the aggregation phase (but not vegetative cells) are attracted to a small peptide. This molecule is either circular or blocked at both ends but its detailed structure is unknown (Wurster, Pan, Tyan & Bonner, 1976). The species *Dictyostelium lacteum* is attracted to a negatively charged aromatic heterocyclic compound that is as yet incompletely characterized (Mato, van Haastert, Krens & Konijn, 1977*a*) and *Dictyostelium minutum* is specifically attracted by a different aromatic moiety bound to a glycine molecule (Kakebeeke, Mato & Konijn, 1978).

Of the attractants, only cAMP used by *D. discoideum* has been studied in detail and the considerable progress made with this system will now be outlined.

Adenylate cyclase and the production and relay of the chemotactic signal. Evidence has been presented that the adenylate cyclase enzyme which produces 3′,5′-cAMP from ATP is located on the inner face of the plasma membrane (Farnham, 1975). More recent work suggests that it may also be contained in vesicles (Hintermann & Parish, 1979). The latter is an interesting finding as apparent fusion of vesicles with the cell surface in phase with cAMP release has been reported by Maeda & Gerisch (1977). It seems possible, although not yet proven, that the cAMP formed may be released from vesicles in the manner of release of neurotransmitters from the preganglionic cells at nerve synapses.

How the adenylate cyclase is activated by starvation is unknown. At some point a few hours after the food supply has been depleted a small number of the cells in a starving population activate their enzyme in an oscillatory manner and thereby produce pulses of cAMP (Roos, Scheidegger & Gerisch, 1977; Klein, 1977). Very few of a normal population of starving amoebae spontaneously produce such pulses of cAMP, but nearly all the cells become excitable and can act as relays for the pulses produced by the cells that do form centres. It may well be that the machinery for relaying cAMP pulses in this way is the same as for initiating pulses spontaneously, and that the steady increase in the excitability of cells as synchronous starvation progresses produces some cells with a self-generating cyclic oscillation of enzyme activity. These cells would then entrain those amoebae around them and inhibit further centre initiation.

Evidence that the adenylate cyclase is briefly activated during relay by external cAMP was provided by Roos & Gerisch (1976) who found that the enzyme was briefly activated in extracts made rapidly after adding a pulse of cAMP to whole amoebae; however, it was not activated when the cAMP pulse was given after the extracts were made. It was concluded that adenylate cyclase activation operates by an indirect system that requires the integrity of the whole cell, and hence implicates binding of the cAMP pulse to the externally facing cell surface cAMP receptors.

The cAMP receptors. The cell surface cAMP receptors that are used in this indirect activation have been extensively studied (Malchow & Gerisch, 1974; Green & Newell, 1975; Mato & Konijn, 1975; Henderson, 1975; Klein & Juliani, 1977; Mullens & Newell, 1978). The properties closely resemble hormone receptors in higher cells and show, for example, curvilinear Scatchard plots indicating multiple classes of receptor or variable receptor affinity dependent on the cAMP concentration. Evidence for such a variable-affinity receptor is the finding that increasing the concentration of cAMP increased the rate of dissociation of previously bound cAMP under conditions that favoured rapid dissociation. The requirement by the organism for such a variable affinity receptor may lie in the need to respond to a wide range of cAMP concentrations such as may be found in the slime moulds' natural habitat.

The number of these receptors has been found to increase rapidly from a basal level during the starvation period with a maximum number present at 8–10 h. After this the receptors disappear equally rapidly and whether they play a role in later development is not clear.

Recently Wallace & Frazier (1979) have used the technique of photoaffinity labelling of amoebae to identify and isolate the cell surface cAMP receptor. The technique employed the compound 8-N_3-c[^{32}P]AMP which binds to cAMP receptors. Subsequent exposure to short-wavelength light causes covalent bonding of the molecules to the receptor protein, which may then be isolated and purified, its presence during purification being monitored by its [^{32}P] label. It was found that the only band on electrophoretic gels that was specifically labelled in this way was a 40 000 Dalton protein with a high specificity for cAMP and a time of formation that identified it as the cAMP receptor.

The linkage system between the cAMP receptor and adenylate cyclase. The connection between the binding of cAMP to the receptor and the activation of adenylate cyclase is an important and interesting problem. From theoretical analysis by Cohen & Robertson (1971) it is clear that there must be a very fast response to binding of cAMP by the cell surface receptors. In order for the relay system to sustain the observed velocity of wave propagation on agar surfaces, the release of cAMP must start within 5-15 s of the binding of the incoming cAMP signal. If this 'relay delay time' were slower than this, either the signal range of each pulse would be very much larger than that observed (about 60 μm) or the wave velocity would be much slower. Slow synthesis of cAMP over a period of minutes has been observed under some conditions (Roos, Nanjundiah, Malchow & Gerisch, 1975; Chung & Coe, 1978) but the physiological significance of such a slow event remains uncertain. Possibly the activation of adenylate cyclase over a period of minutes masks the *release* of cAMP that rapidly occurs following cAMP-receptor stimulation. Evidence that cAMP release is a fast response has been reported by Grutsch & Robertson (1978) who measured the cAMP released from small populations of amoebae on agar in response to pulses of cAMP. They observed that 3×10^7 molecules of cAMP were released by amoebae in response to a stimulus of 100 nM cAMP and that this release was complete within about 8 s.

Cellular events that are observed to be temporarily related over this short time scale (i.e. occur within a few seconds of cAMP binding to receptors) are.

(1) A very rapid increase in intracellular cGMP (Mato *et al.*, 1977*b;* Wurster *et al.*, 1977; Mato & Malchow, 1978);
(2) a very rapid influx of Ca^{++} ions that occurs within a few seconds of cAMP binding followed by a slower efflux over several minutes (Wick, Malchow & Gerisch, 1978); and
(3) a rapid release of protons, about 10^3 protons being released per molecule of cAMP added followed by an increase of the pH to the initial value over about four minutes (Malchow *et al.*, 1978).

Which of these events might represent the link between cAMP receptor stimulation and adenylate cyclase activation is not yet known but all three occur over the approriate time scale.

The chemotactic movement response to cAMP. The change in optical properties that occurs when cells move may also be correlated with cAMP binding. Such changes are best studied quantitatively with the amoebae starving in oxygenated suspension in a spectrophotometer cuvette (Gerisch & Hess, 1974). After addition of cAMP to amoebae that have been starving for a few hours, a rapid decrease in extinction is observed within seconds of cAMP binding. The extinction then rapidly rises towards the original value followed by a slower wave of decreased extinction (Fig. 7). These changes in optical properties may be correlated with those seen on densely populated agar plates and probably represent an initial rapid shock response to the cAMP followed by a slower movement response. This correlation is supported by the observation that streamer mutants, which show a wide movement band on agar, show an extended slow wave of decreased extinction in oxygenated suspensions (Ross & Newell, unpublished).

The way that movement results from the cAMP binding is little understood. However, with Ca^{++}, cGMP and proton changes occurring rapidly after cAMP binding, it may be envisaged that the actomyosin system is activated by changing concentrations of these entities. It is of interest in this connection, that a protein factor that confers Ca^{++} sensitivity to the activation of myosin ATPase by actin has been isolated from *Dictyostelium* amoebae (Mockrin & Spudich, 1976). What is not clear is the way that the direction of movement is established by the amoebae after receipt of a relayed signal from the aggregation centre. In some manner they find their 'chemical compass bearing'. Although this might occur through differential cAMP receptor occupancy from front to back of the amoebae, any cAMP gradient that exists would be very shallow, and it seems more likely that the amoebae are able to recognize which side of a cell was exposed first to a wave of signal reaching it.

Signal destruction: the phosphodiesterases. Unless cAMP were rapidly destroyed after its formation and binding to the cell receptors, the chemosensory

Fig. 7. Oscillations in the extinction of a starving suspension of amoebae in a spectrophotometer cuvette in response to pulses of cAMP. This technique (Gerisch & Hess, 1974) allows small changes in amoebal shape to be monitored that occur in response to cAMP signals in dense suspensions (10^7 to 10^8 amoebae/ml). The amoebae are rapidly mixed by bubbling oxygen through the suspension to one side of the path of the spectrophotometer light beam. Note the sharp decrease in extinction on addition of the cAMP followed by a slower wave of decreased extinction.

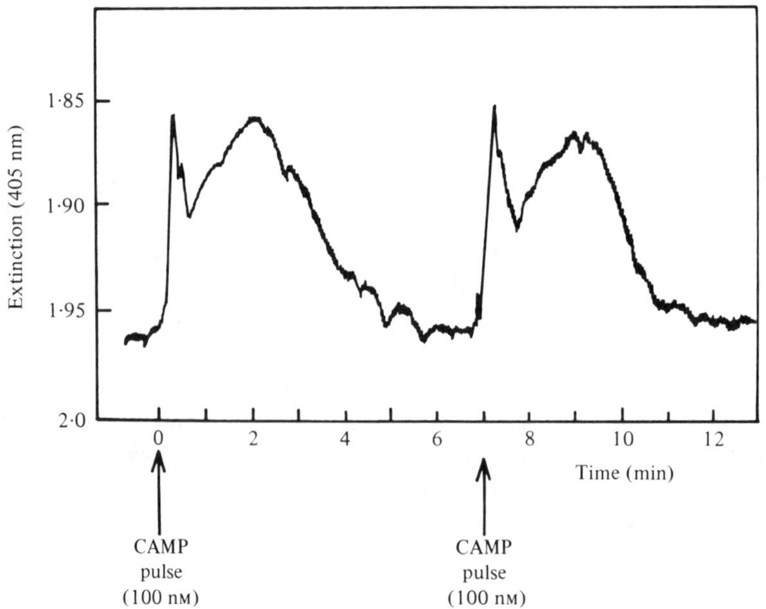

system would soon become swamped and cease to function. In normal cells the cAMP is efficiently removed by the phosphodiesterase enzymes. These enzymes have been much studied and a wealth of literature exists on their varied properties (see reviews by Gerisch & Malchow (1976) and Newell (1977) for references). Briefly stated, there are two major types, one of which is bound to the outside of the plasma membrane and another is excreted as soluble enzyme. A protein inhibitor of the soluble enzyme is also formed and helps to regulate its activity – the inhibitor being repressible by excess cAMP levels (Tsang & Coukell, 1977, 1979; Klein & Darmon, 1977).

Adaptation to cAMP signals. Recent experiments by Devreotes & Steck (1979) have investigated the ability of *D. discoideum* amoebae to adapt to continuously maintained concentrations of cAMP. The experiments employed a specially designed perfusion apparatus in which starving amoebae (prelabelled with [^{3}H]adenosine during the growth phase) were placed on support filters and perfused continuously with buffer containing various concentrations of cAMP. The release of cAMP in response to such external cAMP was monitored by collection of the perfusate and estimation of [^{3}H] by scintillation counting. This technique allowed the effects to be assessed on the perfused amoebae of fixed concentration levels of extracellular cAMP. Devreotes & Steck found that amoebae secrete cAMP only in response to a relative *increase* in the extracellular cAMP concentration. They found, moreover, that the response is graded rather than all-or-none, the size of the relay response being proportional to the increment in extracellular cAMP concentration. The duration of the cAMP secretion depended on the continued presence of extracellular cAMP: removal of the stimulus at any time terminated the response. However, even in the continued presence of the cAMP stimulus, the response ceased after approximately two minutes (Fig. 8). Such cessation of response is indicative of the phenomenon of adaptation that has been extensively studied in the bacterial chemotactic system (Goy & Springer, 1978; see also the chapter in this volume by Hazelbauer). If the perfusing cAMP concentration was carefully increased in tiny incremental steps over a period of 30 min, a continuous secretion of cAMP was found over this period. In any other than this abnormal situation of continuous incremental stimuli, the response is clearly of limited duration due to the adaptation phenomenon. Although the molecular mechanism of adaptation is not yet understood, analogy with the bacterial chemotactic adaptation system suggests that methylation of membrane proteins could play a part. Such methylation has recently been reported in starving amoebae by Mato & Marín-Cao (1979), in response to cAMP signals.

Chemotaxis during the sexual phase

The formation of the zygotes and the resistant macrocyst structure by strains of *D. discoideum* of opposite mating type occurs in response to a pheromone. This volatile substance released by the strain bearing the *matA* allele (e.g. strain NC4) may act non-reciprocally on the strain bearing *mata* (e.g. strain V12) (O'Day & Lewis, 1975; Machac & Bonner, 1975; Lewis & O'Day, 1977). Under appropriate conditions of darkness and in the presence of carbon dioxide, cells of opposite mating type form loose cell clumps. After two of the cells fuse,

Fig. 8. Adaptation to prolonged cAMP stimulii. Amoebae grown in the presence of [^{3}H] adenosine for three hours, were harvested, freed of bacteria and placed on Millipore filters in a perfusion apparatus. After seven hours of dripwise perfusion with buffer, cAMP (1 μM) was dripped on to the filters for periods of up to 30 min as indicated and the perfusate collected and analysed for [^{3}H] cAMP. Note that the liberation of [^{3}H] cAMP ceased after less than five minutes even in the presence of a continuous cAMP stimulus. (Redrawn from Devreotes & Steck (1979).)

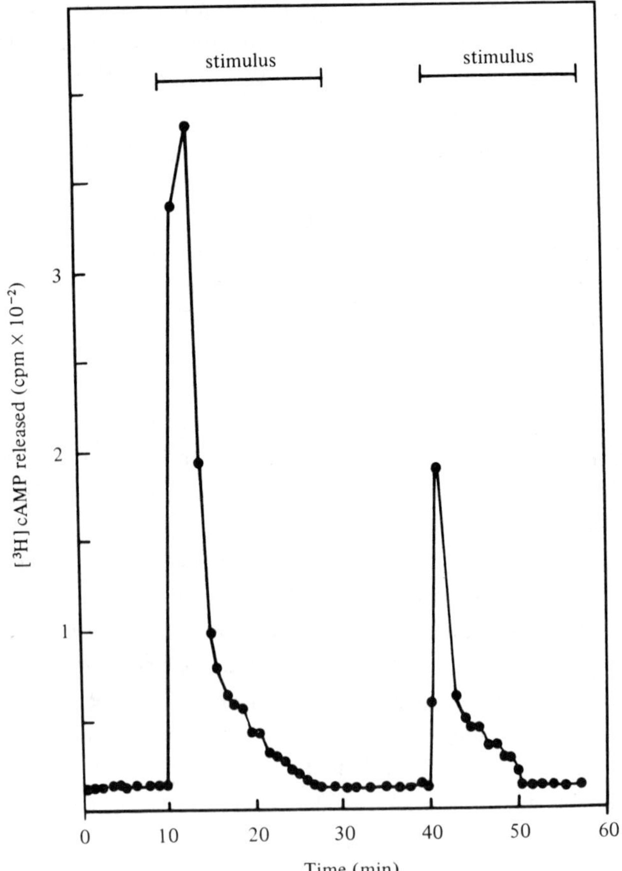

the zygote forms a cytophagic cell which consumes (by engulfment) the mass of surrounding amoebae, simultaneously growing enormously in size. Eventually this giant cytophagic cell becomes surrounded by a thick multilayered wall that matures very slowly (over a period of weeks or months) into a resistant macrocyst. Subsequent germination releases haploid amoebae that have undergone meiosis (see Newell, 1978*a*, for a review).

Recent experiments by O'Day (1979) and O'Day & Durston (1979) have indicated that the formation of the loose cell clumps on agar surfaces may involve chemotaxis to cAMP in a similar manner to chemotaxis during the aggregation phase. The clumps formed are much smaller, however, consisting of only a few hundred cells rather than 20–100000 for aggregates in the normal fruiting pathway. Evidence that chemotaxis is involved is less complete than for aggregation but it appears that the morphology of the process is similar. Streams of cells may be seen with end-to-end contacts moving in a directed manner towards the expanding clumps (Fig. 9). With time-lapse photography, movement in jerks indicative of pulsatile signalling, has also been detected (O'Day, 1979). The clumps themselves can act as chemotactic centres for starving aggregation-competent amoebae, and conversely, amoebae forming macrocyst clumps are able to respond chemotactically to exogenous sources of cAMP placed in the agar. It seems likely, therefore, that chemotaxis on a relatively small scale leads to the formation of the cell clumps which, after consumption by the pheromone-induced cytophagic cells, leads to the formation of the macrocyst.

Fig. 9. Chemotaxis towards clumps of *D. discoideum* amoebae undergoing macrocyst formation. (Reprinted with permission from O'Day (1979).)

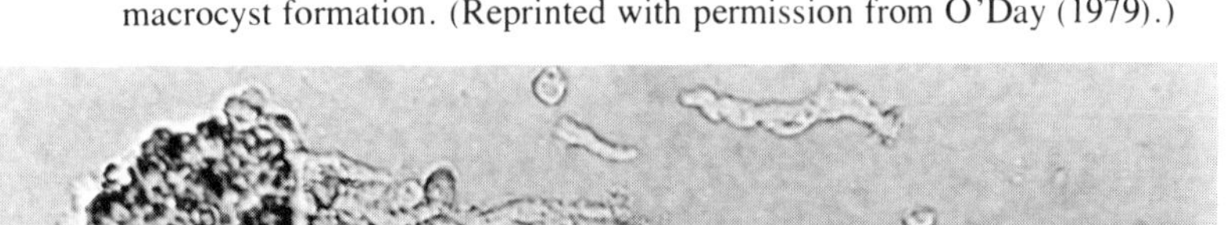

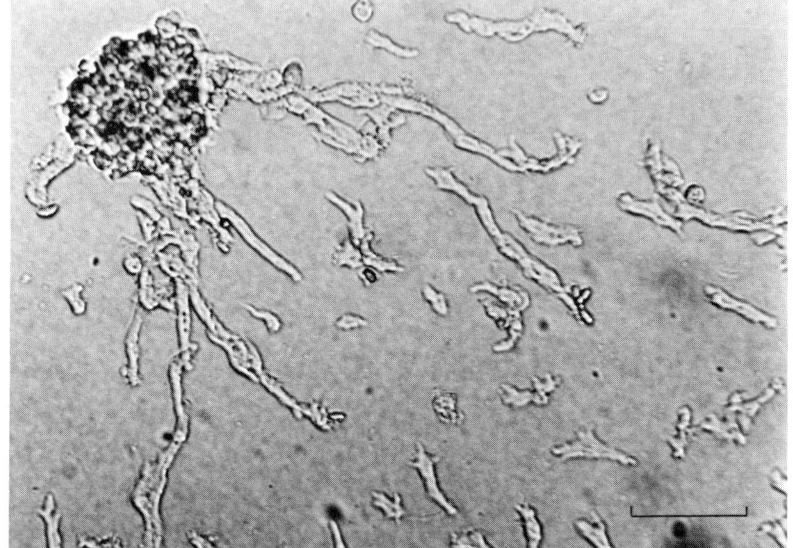

Amoebal chemotaxis in the multicellular aggregate

One of the potentially far-reaching ideas to be found in recent literature about slime moulds concerns the possibility of chemotaxis and pulsatile signalling being regulatory mechanisms operating in the multicellular phase. Stemming mainly from the studies of Durston and coworkers, there is evidence suggesting that chemotaxis does not cease at the stage of aggregate formation but has an important role in regulating the morphological changes seen in the slug. Firstly, it has been observed that slugs and other later structures contain files of cells joined head to tail, rather like those seen in aggregation streams. The files are oriented along the long axis and contain most or all of the cells of the slug. Secondly, slugs and other later structures may often show waves of synchronized cell movements. These show comparable periodicities (2.5–3 min) and velocities to aggregation waves and are non-decremental (i.e. have a roughly constant amplitude as the wave moves down the slug from front to back) (Durston, Vork & Weinberger, 1979) (Fig. 10). Thirdly, paralysis of the cells in certain regions

Fig. 10. Synchronous periodic movements of cells in slugs compared with aggregating amoebae. (*a*) Cells in the posterior zone of a slug. (*b*) Cells in a regenerating slug segment that had the anterior tip removed. (*c*) Anterior cells in a slug of a mutant (KY3). This mutant persists in the slug migration phase and hence is amenable for such analysis. (*d*) Amoebae during the early stages of aggregation, for comparison. (Redrawn from Durston *et al.* (1979).)

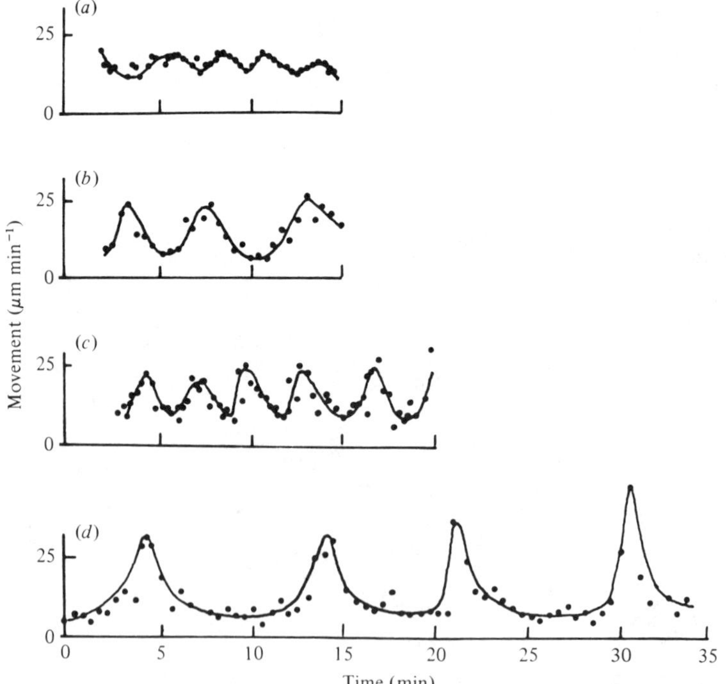

of the slug has been observed that may well account for the shape changes seen as the slugs develop. Slugs contain anterior, middle and posterior zones that are destined to form stalk, spore and basal disc cells respectively (Raper, 1940; Bonner, 1967). Neutral-red staining has been found by Durston *et al.* (1979) to distinguish between these zones because only prestalk and rearguard cells stain well (due to their possessing the autophagic vacuoles that take up the stain (Fig. 11). Using neutral-red staining Durston *et al.* (1979) found that the anterior part of the area destined to become spores (the prespore tissue) was paralysed. The paralysis was made obvious because stained zones next to paralysed tissue remained motile. Simultaneously with such paralysis, propagating waves of movement could be seen in the posterior part of the prespore zone.

The combined action of chemotaxis and paralysis in the slug is postulated by Durston *et al.* to account for the shape and manner of movement of migrating

Fig. 11. Scheme of the transition from a migrating slug to a 'Mexican hat' by a combination of forward chemotactic movement and paralysis. The black areas at the front and rear represent areas stained by neutral-red stain. The stippled area represents the area that becomes paralysed. Cells move forward inside the slime sheath (as indicated by the arrow) during active migration. When anterior cells just behind the tip become paralysed, movement is impeded causing first a swelling in this region and, later on, cessation of movement. It is proposed that the contours of the paralysed zone are distorted by loss of the paralysing substance to the substratum causing the rearmost cells to push under the paralysed zone and form the 'Mexican hat'. (Based on the work of Durston *et al.* (1979).)

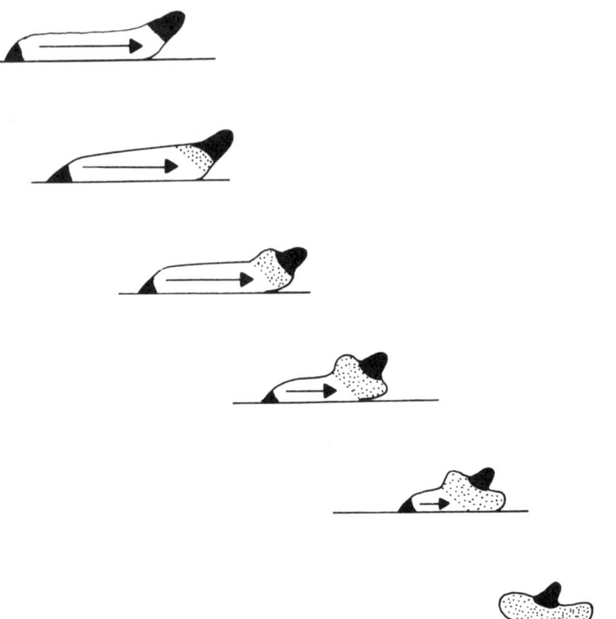

slugs leading to the culmination stage. They consider that the paralysis is caused by a substance (possibly cAMP) secreted by the slug tip (which a few hours earlier was the pulsing aggregation centre) and that this substance only affects prespore cells (or the threshold concentration for prestalk cells is much higher). The paralysed zone would retard movement of slug tissue through the slime sheath and tend to produce the observed widening of the slug just behind the tip. Toward the end of the slug migration period the paralysis is observed to increase and a larger band of the prespore cells becomes motionless. Because of the loss of the paralysing substance to the substratum on which the slug moves, the contours of concentration of the substance would probably be slanted and closer to the tip in the underside of the slug. Durston *et al.* suggest that the increasing concentration of the substance as slug development progresses results in elongation of the paralysed zone until this blocks upper cell movement. As a consequence the tip is pushed upwards by the advancing cells from the rear and comes to the top. At this point the only contact with the substratum is on the base and the rounded slug takes on the symmetrical 'Mexican hat' shape (Fig. 11).

It is also possible that periodic chemotactic movements are again important once culmination is under way and the 'Mexican hat' has started to produce a vertical stalk within the cell mass. Durston *et al.* (1976) report that the increase in length of the stalk may commonly be seen to be by periodic increments, and prestalk cells make periodic waves of movement in synchrony with these increments, due possibly to their movement into the growing stalk. The prespore cells fail to enter the cellulose-encased stalk tube due, presumably, to their remaining paralysed throughout the culmination process.

Signalling by the tip and the consequent responses of either chemotaxis or paralysis may, therefore, account for some of the apparently complex movements of this developing multicellular organism. This is an important notion that deserves more detailed observation not only for explaining the morphogenesis of *Dictyostelium* but also for its implication for other developmental systems.

I wish to thank Mr. Frank Caddick for much help with the photography and for drawing the figures and Mrs. Patricia Stallard for typing the manuscript.

References

Alcântara, F. & Monk, M. (1974). Signal propagation during aggregation in the slime mould *Dictyostelium discoideum*. *Journal of General Microbiology*, **85,** 321–34.

Arndt, A. (1937). Untersuchungen über *Dictyostelium mucoroides* Brefeld. *Wilhelm Roux Archiv für Entwicklungsmechanik der Organismen*, **136,** 681–747.

Barra, J., Barrand, P., Blondelet, M. H. & Brachet, P. (1980). *pds*A: a gene involved in the production of active phosphodiesterase during starvation of

Dictyostelium discoideum amebas. *Molecular and General Genetics,* **177,** 607–13.

Bonner, J. T. (1944). A descriptive study of the development of the slime mold *Dictyostelium discoideum. American Journal of Botany,* **31,** 175–82.

Bonner, J. T. (1949). The demonstration of acrasin in the later stages of the development of the slime mold *Dictyostelium discoideum. Journal of Experimental Zoology,* **110,** 259–71.

Bonner. J. T. (1967). *The Cellular Slime Molds,* 2nd edn. Princeton: Princeton University Press.

Bonner, J. T. (1977). Some aspects of chemotaxis using cellular slime molds as an example. *Mycologia,* **69,** 443–59.

Brachet, P., Dicou, E. L. & Klein, C. (1979). Inhibition of cell differentiation in a phosphodiesterase defective mutant of *Dictyostelium discoideum. Cell Differentiation,* **8,** 255–65.

Chung, W. J. K. & Coe, E. L. (1978). Correlations among responses of suspensions of *Dictyostelium discoideum* to pulses of 3′5′cAMP. *Biochimica et Biophysica Acta,* **544,** 29–44.

Cohen, M. H. & Robertson, A. (1971). Wave propagation in the early stages of aggregation of cellular slime molds. *Journal of Theoretical Biology,* **31,** 101–18.

Coukell, M. B. (1975). Parasexual genetic analysis of aggregation deficient mutants of *Dictyostelium discoideum. Molecular and General Genetics,* **142,** 119–35.

Coukell, M. B. (1977). Evidence against mutational 'Hot-Spots' at aggregation loci in *Dictyostelium discoideum. Molecular and General Genetics,* **151,** 269–73.

Devreotes, P. N. & Steck, T. L. (1979). Cyclic 3′5′ AMP relay in *Dictyostelium discoideum. Journal of Cell Biology,* **80,** 300–9.

Durston, A. J. (1973). *Dictyostelium discoideum* aggregation fields as excitable media. *Journal of Theoretical Biology,* **42,** 483–504.

Durston, A. J. (1974). Pacemaker activity during aggregation in *Dictyostelium discoideum. Developmental Biology,* **37,** 225–35.

Durston, A. J., Cohen, M. H., Drage, D. J., Potel, M. J., Robertson, A. & Wonio, D. (1976). Periodic movements of *Dictyostelium discoideum* sorocarps. *Developmental Biology,* **52,** 173–80.

Durston, A. J., Vork, F. & Weinberger, C. (1979). The control of later morphogenesis by chemotactic signals in *Dictyostelium discoideum.* In *Biophysical and Biochemical Information Transfer in Recognition,* ed. J. G. Vassileva-Popova & E. V. Jensen, pp. 693–708. New York: Plenum Publishing Corp.

Farnham, C. J. M. (1975). Cytochemical localization of adenylate cyclase and 3′5′nucleotide phosphodiesterase in *Dictyostelium discoideum. Experimental Cell Research,* **91,** 36–46.

Franke, J. & Kessin, R. H. (1977). Defined minimal medium for axenic strains of *Dictyostelium discoideum. Proceedings of the National Academy of Sciences of the U.S.A.,* **74,** 2157–61.

Gerisch, G. (1968). Cell aggregation and differentiation in *Dictyostelium. Current Topics in Developmental Biology,* **3,** 157–97.

Gerisch, G. & Hess, B. (1974). cAMP-controlled oscillations in suspended *Dictyostelium* cells. Their relation to morphogenetic cell interactions. *Proceedings of the National Academy of Sciences of the U.S.A.,* **71,** 2118–22.

Gerisch, G. & Malchow, D. (1976). Cyclic AMP receptors and the control of cell aggregation in *Dictyostelium. Advances in Cyclic Nucleotide Research,* **7,** 49–68.

Goy, M. F. & Springer, M. S. (1978). In search of the linkage between receptor and response: The role of a protein methylation reaction in bacterial chemotaxis. In *Receptors and Recognition Series B: Taxis and Behaviour,* ed. G. L. Hazelbauer, pp. 1–34. London: Chapman & Hall.

Green, A. A. & Newell, P. C. (1975). Evidence for the existence of two types of binding site in aggregating cells of *Dictyostelium discoideum. Cell,* **6,** 129–36.

Gross, J. D., Peacey, M. J. & Trevan, D. J. (1976). Signal emission and signal propagation during early aggregation in *Dictyostelium discoideum. Journal of Cell Science,* **22,** 645–6.

Grutsch, J. F. & Robertson, A. (1978). cAMP signal from *Dictyostelium discoideum* amebas. *Developmental Biology,* **66,** 285–93.

Henderson, E. J. (1975). The cyclic adenosine 3′5′monophosphate receptor of *Dictyostelium discoideum. Journal of Biological Chemistry,* **250,** 4730–6.

Hintermann, R. & Parish, R. W. (1979). The intracellular location of adenylyl cyclase in the cellular slime molds *Dictyostelium discoideum* and *Polysphondylium pallidum. Experimental Cell Research,* **123,** 429–34.

Kakebeeke, P. I. J., de Wit, R. J. W., Kohtz, S. D. & Konijn, T. M. (1979). Negative chemotaxis in *Dictyostelium* and *Polysphondylium. Experimental Cell Research,* **124,** 429–34.

Kakebeeke, P. I. J., de Wit, R. J. W. & Konijn, T. M. (1980). A novel chemotaxis regulating enzyme that splits folic acid into 6-hydroxymethylpterin and *p*-aminobenzoyl-glutamic acid. *FEBS Letters, Amsterdam,* **115,** 216–20.

Kakebeeke, P. I. J., Mato, J. M. & Konijn, T. M. (1978). Purification and preliminary characterization of an aggregation-sensitive chemoattractant of *Dictyostelium minutum. Journal of Bacteriology,* **133,** 403–5.

Keating, M. T. & Bonner, J. T. (1977). Negative chemotaxis in cellular slime molds. *Journal of Bacteriology,* **130,** 144–7.

Klein, C. (1977). Changes in adenylate cyclase during differentiation of *Dictyostelium discoideum. FEMS Microbiology Letters,* **1,** 17–19.

Klein, C. & Darmon, M. (1977). Effects of cAMP pulses on adenylate cyclase and the phosphodiesterase inhibitor of *Dictyostelium discoideum. Nature, London,* **268,** 76–8.

Klein, C. & Juliani, M. H. (1977). cAMP-induced changes in cAMP binding sites on *Dictyostelium discoideum* amoebae. *Cell,* **10,** 329–35.

Konijn, T. M. (1970). Microbiological assay of cyclic AMP. *Experientia,* **26,** 367–9.

Konijn, T. M. (1972). Cyclic AMP as a first messenger. *Advances in Cyclic Nucleotide Research,* **1,** 17–31.

Lewis, K. E. & O'Day, D. H. (1977). Sex hormone of *Dictyostelium discoideum* is volatile. *Nature, London,* **268,** 730–1.

Machac, M. A. & Bonner, J. T. (1975). Evidence for a sex hormone in *Dictyostelium discoideum. Journal of Bacteriology,* **124,** 1624–5.

Maeda, J. & Gerisch, G. (1977). Vesicle formation in *Dictyostelium discoideum* cells during oscillations of cAMP synthesis and release. *Experimental Cell Research,* **110,** 119–26.

Malchow, D. & Gerisch, G. (1974). Short-term binding and hydrolysis of cyclic 3′:5′-adenosine monophosphate by aggregating *Dictyostelium* cells. *Proceedings of the National Academy of Sciences of the U.S.A.,* **71,** 2423–7.

Malchow, D., Nanjundiah, V., Wurster, B., Eckstein, F. & Gerisch, G. (1978). cAMP-induced pH changes in *Dictyostelium discoideum* and their control by Ca^{++}. *Biochimica et Biophysica Acta,* **538,** 473–80.

Mato, J. M. & Konijn, T. M. (1975). Chemotaxis and binding of cyclic AMP in cellular slime molds. *Biochimica et Biophysica Acta,* **385,** 173–9.

Mato, J. M., Krens, F. A., van Haastert, P. J. & Konijn, T. M. (1977). 3′5′cAMP-dependent, 3′5′cGMP accumulation in *Dictyostelium discoideum. Proceedings of the National Academy of Sciences of the U.S.A.,* **74,** 2348–51.

Mato, J. M. & Malchow, D. (1978). Guanylate cyclase activation in response to chemotactic stimulation in *Dictyostelium discoideum. FEBS Letters, Amsterdam,* **90,** 119–22.

Mato, J. M. & Marín-Cao, D. (1979). Protein and phospholipid methylation during chemotaxis in *Dictyostelium discoideum* and its relationship to calcium movements. *Proceedings of the National Academy of Sciences of the U.S.A.,* **76,** 6106–9.

Mato, J. M., van Haastert, P. J. M., Krens, F. A. & Konijn, T. M. (1977*a*). An acrasin-like attractant from yeast extract specific for *Dictyostelium lacteum. Developmental Biology,* **57,** 450–3.

Mato, J. M., van Haastert, P. J., Krens, F. A., Rhijnsburger, E. H., Dobbe, F. C. P. M. & Konijn, T. M. (1977*b*). Cyclic AMP and folic acid mediated cyclic GMP accumulation in *Dictyostelium discoideum. FEBS Letters, Amsterdam,* **79,** 331–6.

Mockrin, S. C. & Spudich, J. A. (1976). Calcium control of actin activated myosin ATP-ase from *Dictyostelium discoideum. Proceedings of the National Academy of Sciences of the U.S.A.,* **73,** 2321–5.

Mullens, I. A. & Newell, P. C. (1978). cAMP binding to cell surface receptors of *Dictyostelium. Differentiation,* **10,** 171–6.

Newell, P. C. (1977). Aggregation and cell surface receptors in cellular slime moulds. In *Receptors and Recognition Series B: Microbial Interactions,* ed. J. L. Reissig, pp. 1–57. London: Chapman & Hall.

Newell, P. C. (1978*a*). Cellular communication during aggregation of *Dictyostelium.* The second Fleming Lecture. *Journal of General Microbiology,* **104,** 1–13.

Newell, P. C. (1978*b*). Genetics of the cellular slime molds. *Annual Review of Genetics,* **12,** 69–93.

O'Day, D. H. (1979). Aggregation during sexual development of *Dictyostelium discoideum. Canadian Journal of Microbiology,* **25,** 1416–26.

O'Day, D. H. & Durston, A. J. (1979). Evidence for chemotaxis during sexual development in *Dictyostelium discoideum. Canadian Journal of Microbiology,* **25,** 542–4.

O'Day, D. H. & Lewis, K. E. (1975). Diffusible mating-type factors induce macrocyst development in *Dictyostelium discoideum. Nature, London,* **254,** 431–2.

Pan, P., Hall, E. M. & Bonner, J. T. (1972). The bacterial attracting substance is folic acid. *Nature, New Biology,* **237,** 181–2.

Pan, P., Hall, E. M. & Bonner, J. T. (1975). Determination of active portion of folic acid molecule in CSM chemotaxis. *Journal of Bacteriology,* **122,** 185–91.

Pan, P. & Wurster, B. (1978). Inactivation of chemoattractant folic acid by cellular slime molds and identification of reaction product. *Journal of Bacteriology,* **136,** 955–9.

Raper, K. B. (1940). Pseudoplasmodium formation and organisation in *Dictyostelium discoideum. Journal of the Elisha Mitchell Scientific Society,* **56,** 241–82.

Roos, W. & Gerisch, G. (1976). Receptor mediated adenylate cyclase activation in *Dictyostelium discoideum. FEBS Letters, Amsterdam,* **68,** 170–2.

Roos, W., Nanjundiah, V., Malchow, D. & Gerisch, G. (1975). Amplification of cyclic AMP signals in aggregating cells of *Dictyostelium discoideum. FEBS Letters, Amsterdam,* **53,** 139–42.

Roos, W., Scheidegger, C. & Gerisch, G. (1977). Adenylate cyclase activity oscillation as signals for cell-aggregation in *Dictyostelium discoideum. Nature, London,* **266,** 259–61.

Ross, F. M. & Newell, P. C. (1979). Genetics of aggregation pattern mutations in the cellular slime mould *Dictyostelium discoideum. Journal of General Microbiology,* **115,** 289–300.

Samuel, E. W. (1961). Orientation and rate of locomotion of individual amebas in the life cycle of the cellular slime mold *Dictyostelium mucoroides. Developmental Biology,* **3,** 317–35.

Shaffer, B. M. (1957). Aspects of aggregation in cellular slime molds. *American Naturalist,* **91,** 19–35.

Shaffer, B. M. (1958). Integration in aggregating cellular slime moulds. *Quarterly Journal of Microscopical Science,* **99,** 103–21.

Tsang, A. S. & Coukell, M. B. (1977). The regulation of cAMP phosphodiesterase and its specific inhibitor by cAMP in *Dictyostelium discoideum. Cell Differentiation,* **6,** 75–84.

Tsang, A. S. & Coukell, M. B. (1979). Direct evidence for extracellular adenosine 3′ : 5′-monophosphate phosphodiesterase induction and phosphodiesterase inhibitor repression by exogenous adenosine 3′ : 5′-monophosphate in *Dictyostelium purpureum. European Journal of Biochemistry,* **95,** 419–25.

Wallace, L. J. & Frazier, W. A. (1979). Photoaffinity labelling of cAMP binding and AMP binding proteins of differentiating *Dictyostelium discoideum* cells. *Proceedings of the National Academy of Sciences of the U.S.A.,* **76,** 4250–4.

Warren, J. A., Warren, W. D. & Cox, E. C. (1975). Genetic complexity of aggregation in the cellular slime mold *Polysphondylium violaceum. Proceedings of the National Academy of Sciences of the U.S.A.,* **72,** 1041–2.

Warren, J. A., Warren, W. D. & Cox, E. C. (1976). Genetic and morphological study of aggregation in the cellular slime mold *Polysphondylium violaceum. Genetics, Princeton,* **83,** 25–47.

Wick, U., Malchow, D. & Gerisch, G. (1978). cAMP stimulated Ca^{++} influx into aggregating cells of *Dictyostelium discoideum. Cell Biology International Reports,* **2,** 71–9.

Williams, K. L. & Newell, P. C. (1976). A genetic study of aggregation in the cellular slime mould *Dictyostelium discoideum* using complementation analysis. *Genetics, Princeton,* **82,** 287–307.

Wurster, B., Pan, P., Tyan, G. G. & Bonner, J. T. (1976). Preliminary characterization of acrasin of the cellular slime mold *Polysphondylium violaceum. Proceedings of the National Academy of Sciences of the U.S.A.,* **73,** 795–9.

Wurster, B., Schubiger, K., Wick, U. & Gerisch, G. (1977). Cyclic GMP in *Dictyostelium discoideum:* oscillations and pulses in response to folic acid and cAMP signals. *FEBS Letters, Amsterdam,* **76,** 141–4.

G.W.GOODAY

Chemotaxis in the eukaryotic microbes

The extent of chemotaxis

The organisms to be considered here display an enormous variety of structures, life cycles and habitats. They are the motile cells of the fungi, algae, protozoa and the acellular slime moulds, myxomycetes. The cellular slime moulds are dealt with in detail by P. C. Newell in the preceeding chapter. What all these different cells have in common is that their most important interactions with their environments are chemical – their major senses are smell and taste. They are surrounded by continually changing clouds of an infinite variety of chemicals, and from this milieu they pick out information telling them where they are. Their chemotactic responses are thus the result of their using this information to move.

It is probable that all of the many thousands of species of eukaryotic microbes show chemosensory responses, but only a few cases have been investigated in detail. Some of these motile microbes are of great economic and social importance to man, notably the Protozoa pathogenic to man and animals, and fungi, such as *Phytophthora* species, pathogenic to plants. Most of the work discussed here, however, has used organisms that are more amenable to laboratory manipulation, such as the protozoan *Paramecium tetraaurelia,* the alga *Chlamydomonas reinhardii,* and the myxomycete *Physarum polycephalum*. They are our 'pets', whose life styles fit in with ours. Although the use of these organisms may be justified by seeing them as 'model systems' for the eventual elucidation of sensory mechanisms in 'higher organisms', they clearly are of great interest in their own right.

Chemosensory behaviour in the microbes serves to maintain them in, or guide them into, a favourable chemical environment. From an ecological point of view, such responses can be classified most conveniently into those where the cell is responding to some ubiquitous chemical, which it is likely to find in its environment, and those where it is responding to a unique metabolite from another living cell. In the former group come the feeding responses of cells to potential sources of nutrients, and the avoidance responses to potentially noxious chemicals; in the latter come the remarkably sensitive and specific sexual attractions between gametes, aggregation responses such as that of *Dictyostelium discoideum,* and in some cases perhaps specific attractions of pathogens or symbionts to their hosts. In addition there are responses of cells to alien xenobiotic chemicals, such as drugs and antimetabolites, that are considered only where they throw light on the mechanisms involved in chemotaxis.

Two semantic digressions

The term chemotaxis is used here in its broadest sense to denote the net directional response of a cell to a chemical in its environment. This use does not imply the mechanism of the response. As these mechanisms become elucidated, however, they can be classified with more precision by the use of further terms (Carlile, 1975, 1980*a,b*). Thus a kinesis is a change in speed in response to a chemical stimulus (orthokinesis) or a change in frequency of change of direction (klinokinesis). Chemotaxis in bacteria (see the chapter by G. Hazelbauer) is thus classified as a klinokinesis. Topotaxis is the orientation of an organism by its spatial sensing of stimulus intensity at different points of its body surface. These different mechanisms are not mutually exclusive, and so an organism may show simultaneous changes in speed, angle of turn and orientation in response to a chemical.

The phylogenetic relationships between many of the eukaryotic microbes remain obscure, and this has resulted in territorial disputes among biologists. Thus mycologists and protozoologists both lay claim to slime moulds, the phycologists and protozoologists both lay claim to the flagellated algae, and the rumen protozoan *Neocallimastix frontalis* is the zoospore of a phycomycete fungus. All of these organisms can be studied in parallel however by microbiologists, by grace of the five-kingdom classification (see Margulis, 1970) where they are brought together in the kingdom Protista, instead of being split arbitrarily between the plant and animal kingdoms.

Assays of chemotaxis

Assays for chemotaxis have to be designed with great care. At one extreme for example, a toxin in a capillary tube might give the appearance of an attractant to the unwary, as cells will accumulate in the tube as they are immobilized. For any particular chemical, a range of concentrations has to be used, as it may act as attractant and repellent at different concentrations. As stressed by Bean (1979), whatever assay is used, great attention must be paid to the time course of attraction. Use of capillary tubes might be complicated by other responses of the cells, such as geotaxis (with the cells unable to escape), phototaxis (with the tube acting as a light guide), or responses to differences in aeration in the tube or to the disturbance of the chemical gradient by the swimming of the cells themselves.

With appropriate controls however, the capillary tube assay, basically as used by W. Pfeffer a century ago, can be very useful. It can be used for 'swim-in' tests, with the cells initially in the outside medium (e.g. Sjoblad, Chet & Mitchell, 1978), or for 'swim-out' tests, with the cells initially in the tube (e.g. Allen & Newhook, 1973). In either case, after an appropriate time the number of cells in the tube is estimated: by a direct microscopic count, by blowing out and counting in a particle counter or by plating on an agar medium, or in the case of prelabelled cells by counting their radioactivity (Hirschberg & Rodgers, 1978). Allen & Harvey (1974) observed the speed at which a distinct 'front' of cells advanced into the tube as zoospores of *Phytophthora cinnamomi* retreated from repellents. Modifications of the classical capillary tube assays include filling the tube with agar containing the chemical under investigation and the use of a T-maze, in which the cells have the choice of swimming into the tube containing the chemical or a control tube (Van Houten, 1977).

Demonstrations of positive chemotaxis can sometimes be made by observing the behaviour of swimming cells to agar blocks, particles of plastic, silica or other adsorbent, or oil droplets containing the chemicals (e.g., Müller, 1976*a*). Pommerville (1977) has used an agar plate with wells cut in it to study chemotaxis of gametes of *Allomyces macrogynus*. Test solutions or cells were placed in the centre well, and the response of gametes were observed in the peripheral wells.

Some cells settle or encyst on surfaces in response to the attractant. This phenomenon has been used by Machlis (1973), who assayed the sex attractant sirenin by counting the number of male gametes settling on a cellophane membrane separating them from a chamber containing the test solution, and by Khew & Zentmeyer (1973) who counted the number of zoospores of *Phytophthora* species that encysted around the mouths of agar-filled capillary tubes.

Photographic techniques, for example using darkfield illumination with timed exposures or with stroboscopic lighting, are proving invaluable in elucidating precise responses of swimming cells to chemicals (e.g. Pommerville, 1978; Dryl, 1973).

Chemotaxis of myxomycete plasmodia

These are remarkable organisms, growing in their feeding phase as coenocytic multinucleate plasmodia crawling over surfaces, with internal organizations of pulsating veins channelling protoplasm back and forth. Their behaviour provided some of the first observations of chemotaxis a century ago, when E. Stahl and W. Pfeffer investigated streaming of plasmodia towards and away from a range of substances placed near their peripheries. Elie Metchnikoff described further experiments on the myxomycete plasmodium, 'the colossal amoeboid organism' that 'offers many advantages for the study of protoplasm in general and of pathological phenomena in particular', in his 'Lectures on the Comparative Pathology of Inflammation' given at the Pasteur Institute in 1891. He used his observations on the responses of plasmodia – the positive chemotaxis to substances such as quinine, the digestion of engulfed bacteria but the ejection of a thorn – as a model in the development of his theory of the fundamental nature of the inflammatory reactions in man, which has led to our current understanding of the mechanisms of inflammation and immunological response.

In nature, the plasmodium crawls over substrates, devouring bacteria and fungi. A favorite food of *Badhamia utricularis* is the fruit body of the small bracket fungus *Stereum hirsutum*. Madelin, Audus & Knowles (1975) have shown that small pieces or aqueous extracts of these fruit bodies strongly attract the plasmodium when tested in quadrant plates. In this test, the material is placed in the centre of one-quarter of the agar in a Petri dish, the other three-quarters act as controls, and a plasmodium on a piece of agar is put into the centre of the plate. The direction of movement of the plasmodium is scored. After further studies, Madelin *et al.* ascribed this attraction to a specific highly active compound that is sufficiently volatile to attract plasmodia via an air gap as well as by diffusing through the medium.

Using a similar quadrant test, Knowles & Carlile (1978) found that plasmodia of *Physarum polycephalum* migrated towards wells containing a range of carbohydrates: glucose, galactose, mannose, maltose, *N*-acetylglucosamine, and mannitol (Fig. 1). These attractive carbohydrates are all metabolized and support growth of *P. polycephalum,* whereas sucrose, sorbose and methyl-α-glucoside do not and were not attractants. Ribose and fructose support growth but were not attractants. Acknowledging that the quadrant test depends on diffusion from the

Fig. 1. Chemotaxis of a plasmodium of *P. polycephalum* in the quadrant test (from Knowles & Carlile, 1978). The top well contains 100 mM glucose (250 μl); the others are controls. The plasmodium, on a 1.3-cm membrane filter disc, was inoculated in the centre of the 9-cm plate of agar (salts, vitamins, peptone) and incubated at 24°C for (*a*) 15.5 and (*b*) 20 h. The morphology of the plasmodium indicates that an approximately uniform spreading has been followed by oriented movement towards the test well when the attractant was encountered.

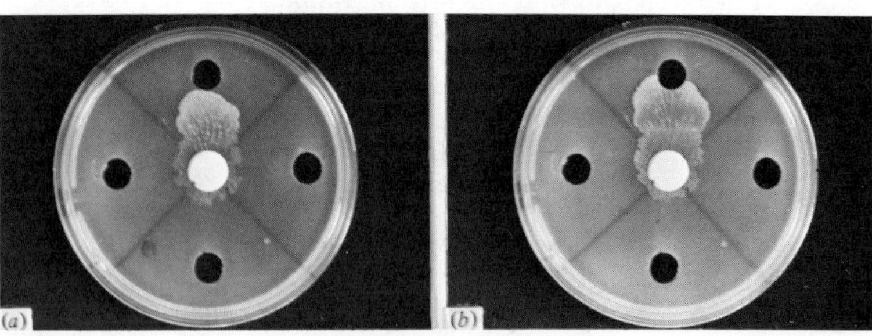

well of the chemical, with consequent uncertainties of concentrations in space and time in the agar, Knowles & Carlile designed the simple technique of the 'double strip test' (Fig. 2). In this system the plasmodium is given the choice of migration onto one of two agar strips, one or both of which contain a uniform concentration of chemicals under test. Using these double strips, the qualitative observations obtained with the quadrant test were confirmed, but in addition, 2-deoxyglucose, which is not metabolized, acted as an attractant. Threshold concentrations for attraction were obtained (Table 1). Negative chemotaxis, repulsion, was observed at high concentrations of the carbohydrates, perhaps in

Table 1. *Chemotactic responses of plasmodia of* Physarum polycephalum *to carbohydrates in the 'double strip test'* [a]

Carbohydrate	Metabolized by plasmodium?	Chemotactic response [b]	Threshold conc. (mM) [c]
Ribose	Yes	None	—
Glucose	Yes	Att	0.25
Galactose	Yes	Att	0.5
Mannose	Yes	Att	5
Fructose	Yes	None	—
Sorbose	No	None	—
Maltose	Yes	Att	0.5
Sucrose	No	None	—
N-Acetylglucosamine	Yes	Att	1
Mannitol	Yes	Att	1
2-Deoxyglucose	No	Att	0.5
Methyl-α-glucoside	No	None	—

[a] Compiled from Knowles & Carlile (1978).
[b] To 10 mM; Att, attraction; None, random migration.
[c] Lowest concentration at which significant attraction occurred.

Fig. 2. Chemotaxis of plasmodia of *Physarum polycephalum* in the double strip test (from Knowles & Carlile, 1978). Two strips of mineral salts agar (20 × 1.6 × 0.5 cm) are three millimetres apart on a glass slide. The top strip contains 10 mM glucose. Plasmodia on 1.1-cm membrane filter discs were placed astride the gap, and incubated at 24°C for 4 h. All plasmodia have migrated towards the glucose agar.

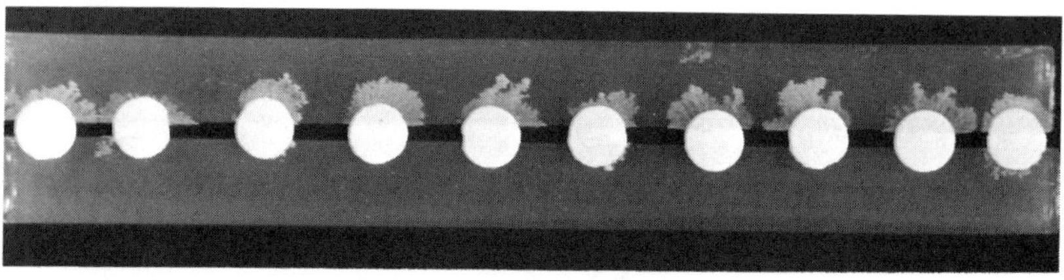

response to the high osmotic pressure. The specificity of the chemoreception was examined by competition, in which the carbohydrates were tested in one agar strip against a background of the same or other chemicals in both strips. The attractants fell into three classes: glucose, galactose, mannose, maltose and 2-deoxyglucose competed with each other; *N*-acetylglucosamine did not compete with, and was not inhibited by, any other; mannitol did not interfere with responses to other attractants, but was not attractive in the presence of carbohydrates in the first two classes. Knowles & Carlile suggest that the positive chemotaxis results from the biassing of the back-and-forth streaming of the protoplasm towards the point of application of the attractant. Ueda & Kobatake (1977) have measured differences in hydrostatic pressure and electrical potential between two portions of a plasmodium of *P. polycephalum* when one portion is stimulated by the addition of chemical. They used a range of organic acids, alcohols and aldehydes. They equated changes in these two measurements to a chemotactic response and a change in membrane potential respectively and found that the threshold concentrations for responses were similar for both.

When starved, the plasmodia of myxomycetes produce sporing structures, analogous to those of the cellular slime moulds (see the preceding chapter by P. C. Newell). The spores germinate to give myxamoebae. These are attracted to food sources of bacteria, and to bacterial extracts, just as are those of *D. discoideum*. The attractant has not been investigated in detail, but it is not cAMP.

Chemotaxis of plant pathogenic zoospores

Members of the Oomycotina, *Pythium* and *Phytophthora* species, are very important plant pathogens, causing diseases of many crop plants and trees. For example, *Phytophthora infestans* causes potato blight, and was responsible for the Irish potato famine of the last century. The propagation of many of these diseases involves the production of zoospores which swim through water films and through wet soils to infect new plants. The zoospores are attracted by roots and root exudates, and this chemoresponse presumably is a major factor in the spread of disease.

Research has taken two lines: identification of attracting molecules in root exudates; and testing of pure chemicals as attractants (Gooday, in Carlile, 1975). Root exudates have proved to be complex mixtures containing a wide range of metabolites. Some components, such as amino acids and sugars, act as attractants (Khew & Zentmeyer, 1973). Fig. 3 illustrates the chemotaxis of zoospores of *Phytophthora palmivora* to dansyl-asparagine, a fluorescent derivative of the amino acid. These metabolites are unlikely to be major natural attractants, however, as they are present in exudates in concentrations which are too low, and the pure compounds and mixtures of them are not such powerful attractants as the exudates themselves.

Allen & Newhook (1973) have suggested that ethanol might be an important natural attractant, being a metabolite of roots in partially anaerobic conditions such as waterlogged soil. They demonstrated that zoospores of *P. cinnamomi* are positively chemotactic to ethanol solutions (Table 2). In the 'swim-in' test, the number of zoospores was counted that accumulated in the capillary tube

Table 2. *Chemotaxis of* Phytophthora cinnamomi *zoospores to ethanol* [a]

Test	Number of zoospores in capillary tube	
	Control	5 mM ethanol
'Swim-in', ethanol in tube, 30 min	22	810
'Swim-out', water in tube, 30 min	42	18
'Rhizosphere model', conc. gradient over 16 mm, 2 h	15	83
'Ideal soil', static, 10 min	7	53
'Ideal soil', slow flow, 10 min	8	28
'Ideal soil', fast flow, 10 min	2	5

[a] Compiled from Allen & Newhook (1973) and Young, Newhook & Allen (1979).

Fig. 3. Chemotaxis of zoospores of *P. palmivora* to the fluorescent derivative, dansyl-asparagine. A 1-μl capillary tube (internal diameter 190 μm) containing 10 mM dansyl-L-asparagine was placed in a pool of zoospore suspension. Photographs were taken (*a*) with transmission fluorescence optics (excitation filter BG 12, barrier filter K510) to visualize the gradient of yellow fluorescent chemical emanating from the tube, and (*b*) with darkfield white light to show the cluster of zoospores around the mouth of the tube. (Photographs kindly provided by J. N. Cameron and M. J. Carlile.)

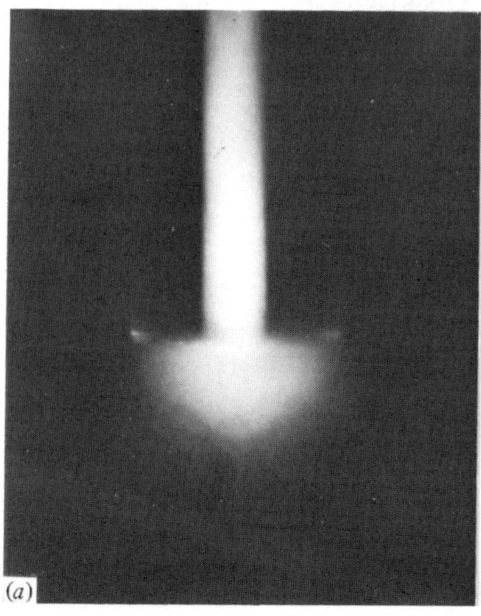

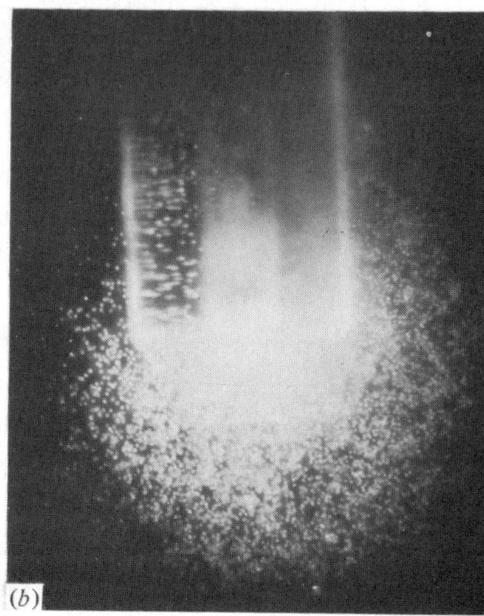

(190 μm in diameter) containing either ethanol solution or water. Tactic responses were observed in the first three to five minutes, and the swimming cells showed a series of turning reactions to move towards the source of the ethanol. In the 'swim-out' test, the number of zoospores was counted that remained in the distilled water in a capillary tube which had been placed in an ethanol solution or in a water control. Zoospores in the tube swam towards the source of ethanol at the ends of the tubes. In the 'rhizosophere model', a concentration gradient was established along the capillary tube that acted as a diffusion bridge between a large stirred volume of ethanol solution (or water in the control) and a large stirred volume of water. After equilibration, a stirred zoospore suspension replaced the water, and the number of zoospores entering the tube was counted. The 'ideal soils' (Young, Newhook & Allen, 1979) were horizontal columns of glass beads (bead diameter about 0.7 mm) containing suspensions of zoospores in 1 mM NaC1 solution, either static or flowing ('slow' at 0.38 cm min^{-1}; 'fast' at 5.75 cm min^{-1}). The capillary tubes, containing 5 mM ethanol in 1 mM NaC1, or the saline alone as control, were inserted vertically at intervals along the column and the numbers of zoospores swimming into them were counted. Young *et al.* (1979) conclude that the chemotactic responses to ethanol can occur in particulate systems, and in water flows up to the swimming speeds of the zoospores.

Allen & Newhook (1974) have studied the effect of isotropic solutions of ethanol, from 0 to 25 mM, on swimming characteristics of the zoospores. The ethanol does not affect the speed or angular velocity of the cells (at about 120 μm s^{-1} and 4.3 rad s^{-1} respectively); i.e. the motor activity of the flagella and the hydrodynamic properties of the soma that lead to rotation. The ethanol does, however, suppress the spontaneous turning activity of the cells, which will aid them to swim towards a source of ethanol through confined spaces as in soil.

Cameron & Carlile (1978) have investigated a range of volatile organic compounds, and report that several alcohols, aldehydes and organic acids act as attractants of zoospores of *P. palmivora.* By far the most active compound is isovaleraldehyde, $(CH_3)_2CH \cdot CH_2 \cdot CHO$, the threshold concentration for which is 1 μM in their bioassay, in comparison with 5 mM for ethanol. They suggest that such volatile compounds could act as natural attractants to plant roots. J. N. Cameron & M. J. Carlile (personal communication) have investigated the binding of isovaleraldehyde to the zoospores. They find it to be specific and saturable, with about 1.4×10^5 molecules of isovaleraldehyde bound to each cell. The specificity has been investigated by measuring the abilities of related or unrelated compounds to displace bound radioactively labelled isovaleraldehyde. Only closely related compounds have this ability, and their dissociation constants show a good correlation with their chemotactic abilities. This points to there being specific isovaleraldehyde receptors at the cell surfaces of the zoospores.

Zoospores of *P. cinnamomi* (Allen & Harvey, 1974) and *P. palmivora* (J. N. Cameron & M. J. Carlile, personal communication) show negative chemotaxis to many low-molecular-weight cations. For each cation there is a critical threshold concentration, those for the most active ion, H^+, being 37 and 150 μM respectively for the two species. The effectiveness of repulsion is positively related to the ionic conductivities of the cations. Cameron & Carlile suggest that the cations neutralize some of the negative charges at the zoospore surface, which, as the interior of the cell has a more negative charge than the surface, will result in an increase in transmembrane potential. This hyperpolarization would then result in changed flagellar behaviour and turning of the zoospore.

Chemotaxis of other flagellate cells

The zoospores of the rumen fungus *Neocallimastix frontalis* invade the plant tissue eaten by the host animal, germinate and then grow on it. Using barley awns as bait, Orpin & Bountiff (1978) showed that prior leaching with distilled water reduced the extent of their invasion by zoospores by a hundredfold, suggesting that soluble diffusible materials in the plant tissue were acting as chemotactic agents. By testing a range of likely chemicals they found that zoospores were attracted to carbohydrates but not to amino acids, purines, pyrimidines or vitamins. The chemotaxis assay was a count of the number of zoospores entering a capillary tube. Thus after incubation for an hour, the counts were approximately 7000, 6000, and 2000 zoospores in tubes containing 1, 0.1, and 0.01 mM glucose respectively, in comparison to a control count less than 0.3% of total. Competition experiments, where chemotaxis to a particular carbohydrate in the tube was tested against a background of another carbohydrate in tube and suspension (Table 3) indicated that the zoospores have at least four chemoreceptors for carbohydrates.

(1) The glucose receptor, sensitive to glucose, galactose, xylose, sorbose, fucose and 2-deoxyglucose;
(2) the sucrose receptor, sensitive to sucrose, fructose and raffinose;
(3) the sorbitol receptor, sensitive to sorbitol and mannitol; and
(4) the mannose receptor, sensitive to mannose and glucose.

Threshold concentrations for chemotaxis to single carbohydrates were similar to those reported for bacteria (see the Chapter by G. Hazelbauer), being for example 5μM for glucose and fructose, and 0.1 μM for sucrose. All of the carbohydrates tested that did not elicit chemotaxis, for example ribose, maltose and trehalose, also did not support growth of *N. frontalis*. Some carbohydrates that were attractants, for example xylose and raffinose (Table 3), also did not support growth, and so attractants are not necessarily metabolized. The barley awns were richest in sucrose, glucose and fructose, and low concentrations of mixtures of these three nutrients acted synergistically, giving greater chemotaxis than expected from addition of their separate effects.

The evidence for chemotaxis of zoospores of *Allomyces* species (Chytridiomycotina) and of *Saprolegnia* species (Oomycotina), to nutrients is reviewed by Gooday (in Carlile, 1975).

The flagellate algae, widely studied for their phototaxis, have been much less investigated for their chemotactic responses. Sjoblad *et al.* (1978), however, have studied the chemoresponses of the marine green alga *Dunaliella tertiolecta* to a wide range of inorganic and organic compounds. Only ammonium ion, L-tyrosine, L-tryptophan and L-phenylalanine have proved to be strong attractants, all having threshold concentrations for chemotaxis at about 1 μM. Competition experiments indicate two chemoreceptors, one for ammonium ion, the other for the aromatic amino acids. In a study with *Chlamydomonas reinhardii,* Hirschberg & Rodgers (1978) found chemotaxis only to L-arginine (a repellent, threshold concentration 1 μM) and to Co^{++} and Mn^{++} (attractants, threshold concentrations 0.1 and 1 mM respectively).

Levandowsky & Hauser (1978) present details of chemoresponses of the marine colourless dinoflagellate *Crypthecodinum cohnii,* using as an assay the tendency of the swimming cells to settle and embed themselves into an agar gel. This behaviour may be related to their natural habit of being saprophytic on

Table 3. *Competition experiments to investigate specificity of chemotaxis of zoospores of* N. frontalis *to carbohydrates* [a]

Carbohydrate throughout system (10 mM)	10^{-2} × no. zoospores attracted to carbohydrate in capillary tube			
	Glucose	Sucrose	Sorbitol	Mannose
Glucose	—	68	49	**0.7**
Sucrose	60	—	32	41
Sorbitol [b]	73	69	—	44
Mannose [b]	65	71	50	—
Galactose	**0.8**	69	46	42
Xylose [b]	**0.6**	65	54	47
Sorbose [b]	**0.8**	70	51	45
Fucose [b]	**0.8**	66	48	32
2-Deoxyglucose [b]	**1.2**	72	50	8
Fructose	70	**0.5**	47	38
Raffinose [b]	61	**0.2**	38	43
Mannitol	66	71	**0.2**	49
None	72	76	52	48

[a] Summarized from Orpin & Bountiff (1978). Zoospore suspensions and carbohydrates were in reduced rumen fluid equilibrated with carbon dioxide. Incubation for 1 h at 39°C. Significant competition indicated by bold figures.

[b] These carbohydrates did not support growth.

rotting seaweeds. The cells showed a strong positive response to the presence in the gels of the seaweed metabolites, dimethyl-β-propiothetin and L-fucose, and not to their chemical analogues, dimethyl-acetothetin and D-fucose.

Chemoresponses of *Paramecium* species

Since the classic account by H. S. Jennings in 1906 in his book *Behaviour of Lower Organisms* of the behaviour of paramecium cells, *Paramecium* species have been favourite organisms for study. In recent years they have proved of great value, being especially suitable for genetic studies and for bioelectrical studies using microelectrodes.

Dryl (1963, 1973) describes the marked response of cells of *Paramecium caudatum* to a pH boundary, so that by turning reactions they stay within the area of favoured pH (Fig. 4), and the negative responses to cations in the order of repulsion of Ca^{++}, Na^{+}, Mg^{++}, Li^{+}, K^{+}, Ba^{++}. The avoidance response is klinokinetic, involving a transient reversal of ciliary beat and so a backing away and turning, followed by resumption of normal ciliary beating and so a swimming forward in another direction. Van Houten (1977) has quantified chemotaxis in *P. tetraurelia* by using a T-maze to obtain the index of chemokinesis, I_{che}, the ratio of number of cells in the test solution to that in the control solution. Thus an I_{che} greater than 0.5 indicates attraction and less than 0.5 indicates repulsion. Wild-type cells for example were attracted by acetate, with a value of I_{che} of 0.84 for 5 mM sodium acetate (5 mM NaCl as control). Analysis of this response showed that as expected the frequency of the avoiding reaction decreased in acetate solution relative to the chloride control, i.e. it was consistent with klinokinesis; but the cells swam faster (0.96 mm s^{-1} after 15 min) in the acetate than in the chloride (0.77 mm s^{-1}). This faster swimming in acetate was retained in a mutant (d4-530) with greatly suppressed avoidance frequency, and explained the unexpected repulsion of this mutant by acetate ($I_{che} = 0.14$; conditions as for wild type), i.e. the cells were repelled by swimming faster, by orthokinesis.

These two components of behaviour, avoiding reaction and velocity of forward swimming are both controlled by the membrane potential (Eckert, 1972). Van Houten (1979) has measured directly the membrane potentials with microelectrodes inserted in cells of *P. tetraurelia* before and after changing the chemical composition of the bathing solution. A change to an attractant resulted in a hyperpolarization. Thus changing from 5 mM NaCl to 5 mM sodium acetate led to a change in membrane potential from −27.0 to −37.7 mV. A change to a repellent resulted in a depolarization. Thus changing from 5 mM KCl to 5 mM NaCl (equivalent to an I_{che} of 0.27) led to a change in membrane potential from −31.3 to −27.0 mV. In the former case, the hyperpolarization causes a decrease in frequency of avoiding reaction and, less importantly, an increase in swimming velocity, and so leads to attraction (except for mutant d4-530). In the latter case the depolarization causes an increase in the frequency of the avoiding reaction and a decrease in swimming velocity and so leads to repulsion.

Van Houten points out that for both hyperpolarization and depolarization the factor of swimming velocity will eventually override the factor of frequency of

Fig. 4. Responses of *Paramecium caudatum* to drops of buffer. (*a*) Positive response to buffer at pH 5.6 by cells from medium at pH 8.35. (*b*) Negative response to buffer at pH 8.35 by cells from medium at pH 5.6. Darkfield micrographs, 10-s exposure, with camera diaphragm opened much wider for the last 0.5 s to give a bright spot at the final position of the cell. Scale bar, five millimetres. (From Dryl (1963).)

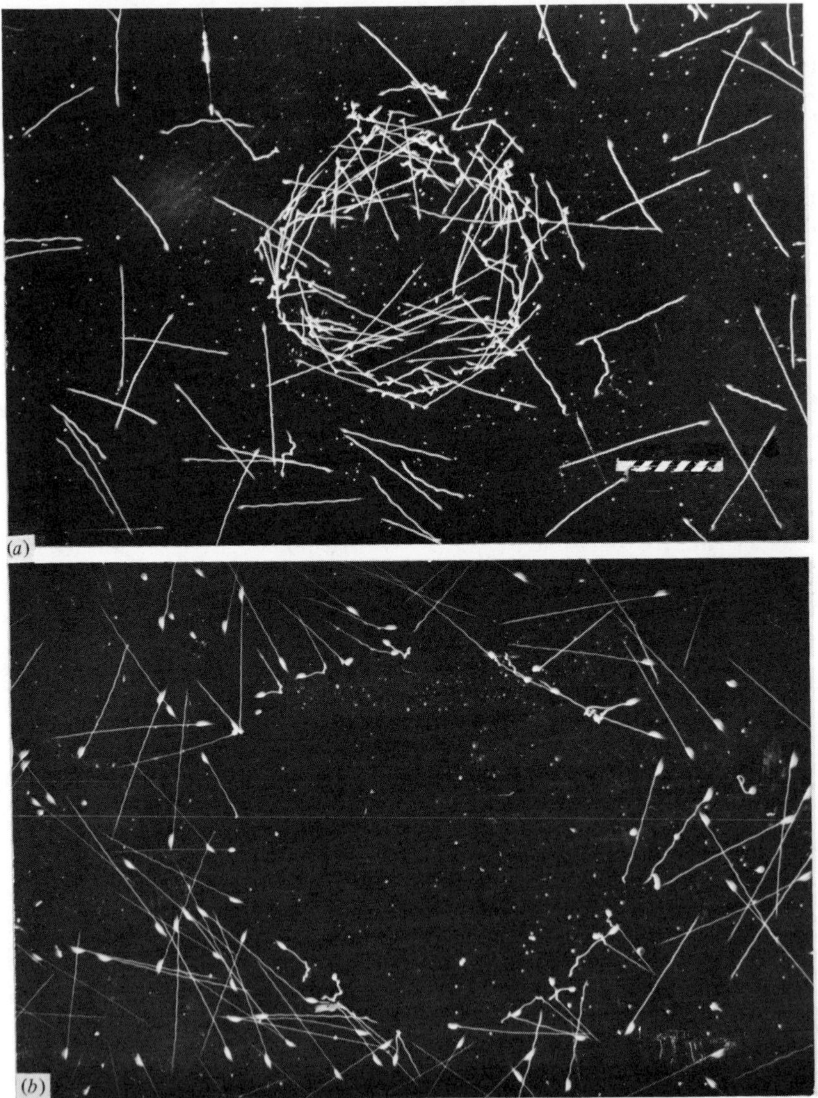

avoiding reaction in determining the response, i.e. klinokinesis will be replaced by orthokinesis. She has found this in two cases. A change from 1 mM KCl to 1 mM KOH (equivalent to repulsion with $I_{che}=0.38$) led to a very large change in membrane potential from −36.8 to −52.3 mV; and a change from 2 mM NaCl to 1 mM $BaCl_2$ (equivalent to attraction with $I_{che}=0.72$) led to a change from −32.9 to −0.9 mV. Thus in the repulsion by KOH, the frequency of avoiding response dropped towards zero, and the cells swam faster and so escaped, while in the attraction by $BaCl_2$, the reverse occurred and the cells were trapped.

Sex attractants

Sexual reproduction is important in allowing genetic recombination. For a microbe, however, the differentiation to produce gametes is a digression from the essential day-to-day activities of growth, cell division and dissemination. It is not surprising therefore to find mechanisms that increase the chances of compatible gametes finding each other amongst cells of the many other species that share their natural environments. One of these mechanisms is the production of, and response to, specific sex attractants. Very few of these substances have been characterized yet, and certainly many await discovery. Only those that have been identified are discussed here.

Sirenin

The sirens of Greek mythology lured sailors to their deaths by their lovely singing, but Odysseus eluded them by filling his crew's ears with beeswax and having himself lashed to the mast of his boat while he sailed past. The sirens of the water mould *Allomyces* are the motile female gametes, that produce the very potent attractant sirenin. This is a bicyclic sesquiterpene, containing a cyclopropane ring (Fig. 5). Its existence was discovered by Leonard Machlis in 1958, and its structure was elucidated ten years later. The bioassay, involving counting the number of male gametes settling on a membrane separating them from the test solution (Machlis, 1973), shows that the male gametes are sensitive to a very wide concentration gradient, from 10^{-10} to 10^{-5} M.

Allomyces species produce five motile cells during their life cycles, haploid and diploid zoospores, male and female gametes and zygotes. Of these, only the male gametes respond to sirenin, and only female gametes produce it. Very quickly after fertilization the motile zygotes develop a positive chemotactic response to amino acids. The chemotactic response of male gametes, down to 22 pg/ml sirenin, is much more sensitive than that of the zygotes, at 400 μg/ml casein hydrolysate (a mixture of amino acids and peptides). Sirenin is rapidly inactivated by responding male gametes, apparently by being metabolized to an inactive product. Pommerville (1977, 1980), using an assay involving the observation of gamete behaviour in wells in agar close to wells containing test materials, suggests that the male gametes also produce a female attractant.

Pommerville (1978) has used darkfield microscopy to analyse the motility of gametes Figs. 6–9; Table 4). The male gametes show a smooth swimming pattern, interrupted by very brief jerks of the cell body resulting in a change of swimming direction. The female gametes swim more slowly, and have a higher frequency of jerks so that they move more sluggishly. Female gametes introduced in the vicinity of males induce an immediate change in their swimming behaviour. The males move in helical paths towards the females, with only a few jerks to reorient themselves. When very near to a female cell, the male undergoes many very short runs and jerks until contact with the female is achieved, resulting in plasmogamy and zygote formation. The swimming pattern

Table 4. *Swimming characteristics of gametes and zygotes of* Allomyces macrogynus [a]

Swimming cell	Velocity (μm s^{-1})	Smooth-swimming distance (μm)	Turn angle	No. of jerks/min
Male gamete	100	50	60°	70
Female gamete	60	65	80°	100
Zygote	140	360	n.d.	n.d.

[a] Compiled from Pommerville (1978). All values are means of at least 10 separate cells, and all differences are significant at $\alpha = 0.01$. n.d., not determined.

Fig. 5. Structures of sex attractants.

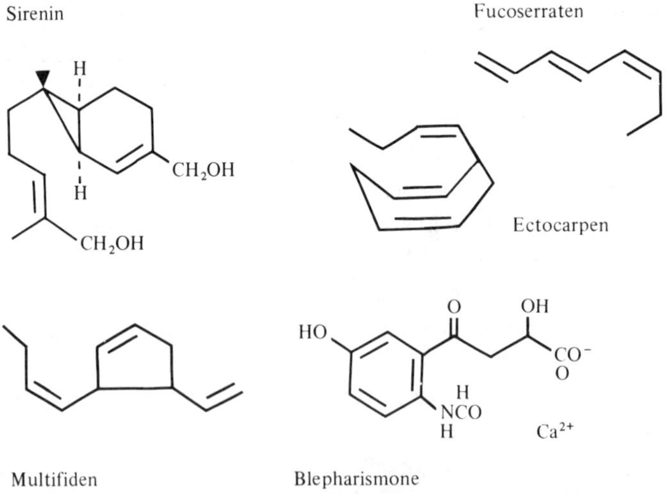

Sirenin

Fucoserraten

Ectocarpen

Multifiden

Blepharismone

of the zygote is distinctive, being faster than that of either gamete, and being in smooth helices involving a continuous turning of the cell body.

The addition of a pulse of sirenin solution to a suspension of male gametes results immediately in rapid jerking movements (Fig. 9(*a*)), like those seen in the immediate vicinity of a female cell (Figs. 8(*b*), (*c*)), with very little net movement in any direction. Over the next 15–30 s, the cells change behaviour to swim in large circles or helices (Fig. 9(*b*),(*c*)), like those seen at some distance from a female cell (Fig. 8(*a*)). In about 60–90 s after the addition of sirenin the cells are reverting to their normal unstimulated swimming patterns, with smooth swimming interspersed with jerks, but some male gametes are seen to swim in clusters (Fig. 9(*d*)), a behaviour never seen in the absence of sirenin.

The sirenin produced by the female gametes thus brings about a very precise orientation and attraction of the male gametes, favouring efficient mating. The

Fig. 6. Random motility tracks of male gametes of *A. macrogynus*. Figs. 6–9 are darkfield micrographs illuminated for set times; starts of swimming track are indicated by arrowheads; all from Pommerville (1978). Fig. 6(*a*),(b) Motility tracks, showing smooth nature of the short runs and the interruptions of these runs by jerks (J) which result in a change of swimming direction. (*a*), 7 s; (*b*), 6 s; both fields of view are 0.28 mm high.
Fig. 7. Random motility tracks of female gametes, showing runs, jerks and changes in swimming direction, but the number of jerks is larger than for the male, so that most females tend to remain in a confined space. Both 6-s exposures; fields of view are 0.31 mm high (*a*), and 0.23 mm high (*b*).

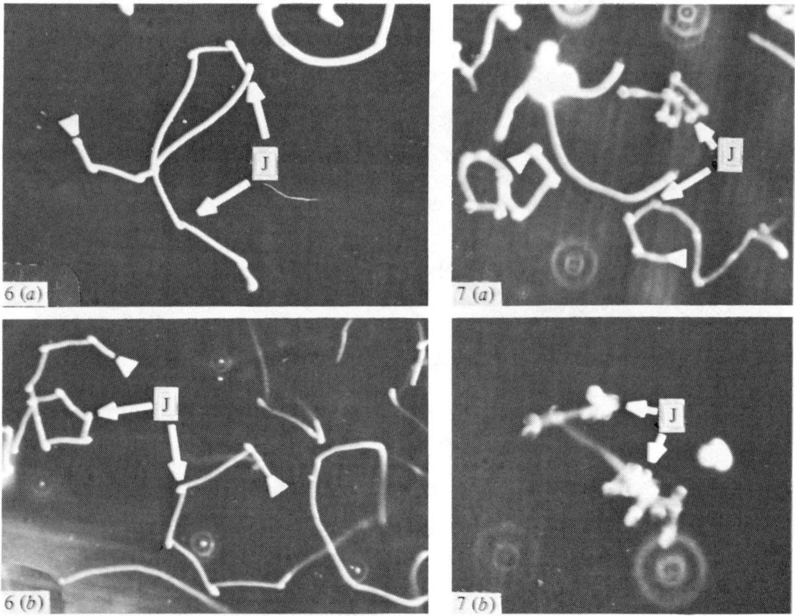

mechanism of sensing the concentration gradient of sirenin could be temporal, with an increase in concentration suppressing the frequency of jerks as a male swims towards a female, and a decrease in concentration increasing the frequency as a male swims away. The change in direction of swimming appears to

Fig. 8. Male chemotaxis of female gametes of *A. macrogynus*. (*a*) Spiral motility of male as he approaches female gamete (FG). Jerks (J) only occur when male cell starts moving away from female. (*b*),(*c*) Spiral motility of males until very near the females, when many jerks occur, resulting in many directional changes until gamete contact occurs. (*a*) 7 s, (*b*),(*c*) 15 s; all fields of view 0.34 mm wide.

Fig. 9. Male gametes responding to sirenin. (*a*) 10 s after the pulse of sirenin. The male cells no longer swim in a random fashion, but undergo numerous jerks resulting in net loss of movement. (*b*), (*c*) 30 and 60 s after pulse, gametes swim in spirals and circles, without jerks. (*d*) 90–120 s after pulse, cells begin to return to normal motility, as shown by presence of jerks (J), but some swim in clusters (MC). (*a*),(*d*), exposure time 4 s; (*b*), 6 s; (*c*), 5 s; (*a*) field of view 0.28 mm high; (*b*), 0.24 mm; (*c*), 0.34 mm; and (*d*) 0.38 mm.

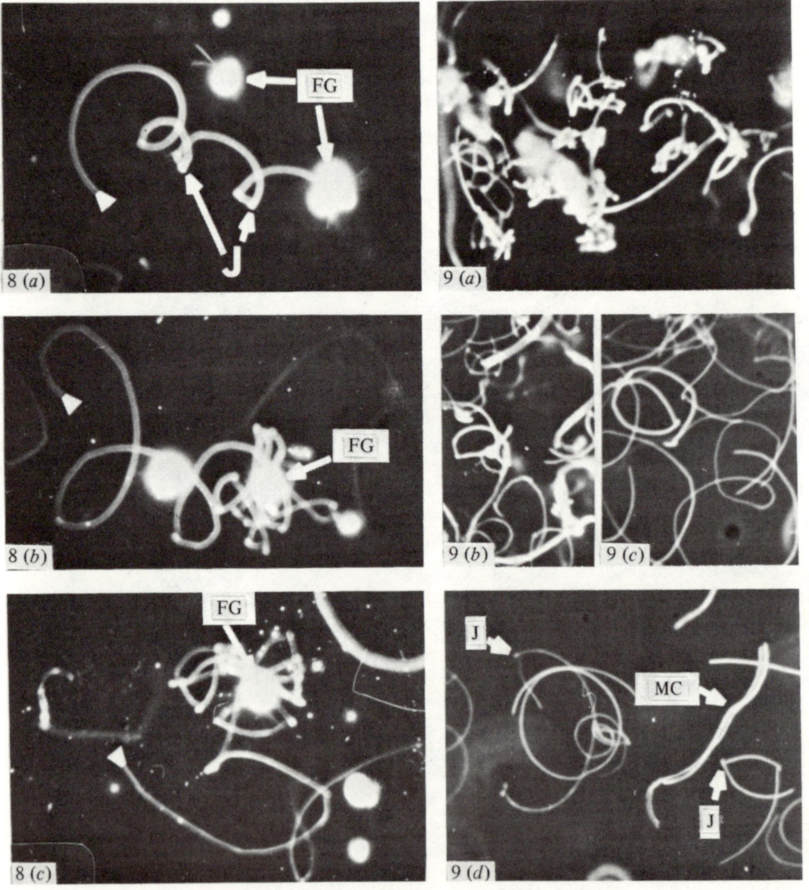

involve the temporary cessation of beating of the flagellum during the 'jerk' (Pommerville, 1978).

Sex attractants of brown algae

The attraction of male gametes to female gametes is a widespread phenomenon in the brown algae, the marine Phaeophyta.

In *Fucus* species, the spermatozoa cluster around the massive eggs (Figs. 10(*a*),(*b*)). In the 1940s, A. H. Cook and J. A. Elvidge used liquid air traps to collect the attractants produced by eggs of *Fucus* species, and attempted to identify them by mass spectrometry. They characterized them as low-molecular-weight hydrocarbons, but it was over twenty years before the attractant for *Fucus serratus* and *Fucus vesiculosus* was identified as *trans, cis*-1,3,5-octatriene, termed fucoserraten (Müller & Seferiadis, 1977; Fig. 5). A yield of 690 μg of fucoserraten was obtained from 252 kg of fresh female receptacles of *F. serratus*. The bioassay consists of exposing a suspension of spermatozoa to the fucoserraten dissolved in the fluorinated solvent FC-78 (Fig. 11), and counting the number of spermatozoa over the droplet after four minutes in comparison to the number over a control droplet of solvent. The spermatozoa are attracted to other hydrocarbons, but less effectively than to fucoserraten. Thus the threshold con-

Fig. 10. Chemotaxis of algal male gametes to female gametes. (*a*) *F. serratus;* (*b*) *F. vesiculosus;* (*c*) *C. multifida*. Darkfield flashlight photographs. Eggs of *Fucus* species are 70 μm in diameter; female gamete of *C. multifida* is 20 μm. ((*a*),(*b*), from Müller & Seferiadas (1977); (*c*), from Müller (1974).)

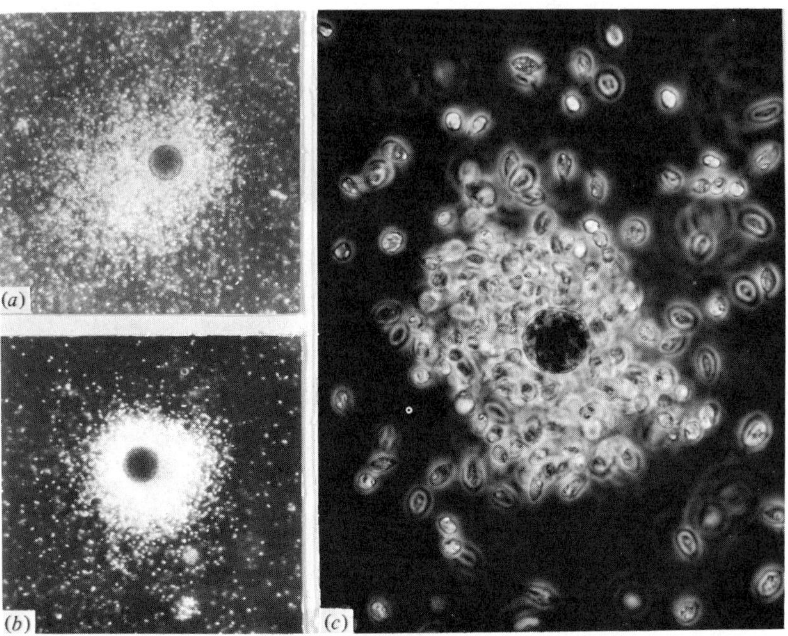

centration in the FC-78 that was detected was 1 μM fucoserraten, but 3.1 mM *n*-hexane. Nevertheless, the lack of total specificity of attractant is sufficient to explain the disruption of sexual reproduction in these algae by coastal oil pollution.

In *Ectocarpus siliculosus* and *Cutleria multifida* the female gametes are motile, being the same size and larger than the motile male gametes respectively, but they soon settle and strongly attract the actively swimming male gametes (Fig. 10). By trapping volatile metabolites from the female algae, Müller (1974, 1976*a,b*) has isolated two specific hydrocarbon attractants, ectocarpen and multifiden (Fig. 5). The bioassays are carried out in the same way as for fucoserraten and *Fucus* gametes, with the attractants being dissolved in droplets of FC-78 (Fig. 11). Again the male gametes are attracted to a range of hydrocarbons, but less effectively than to their natural attractant. Thus threshold concentrations in

Fig. 11. Chemotaxis of algal male gametes to sex attractants in droplets of the solvent FC-78. (*a*) *F. serratus* spermatozoa around droplet containing 100 μM fucoserraten; (*b*) *C. multifida* spermatozoa and a control droplet of FC-78 (left) and a droplet containing 31 μM multifiden (right). Darkfield flashlight photographs. Droplets are 0.4 mm (*a*) and 0.9 mm (*b*) in diameter. ((*a*), from Müller & Seferiadis (1977); (*b*), from Müller (1976*a*).)

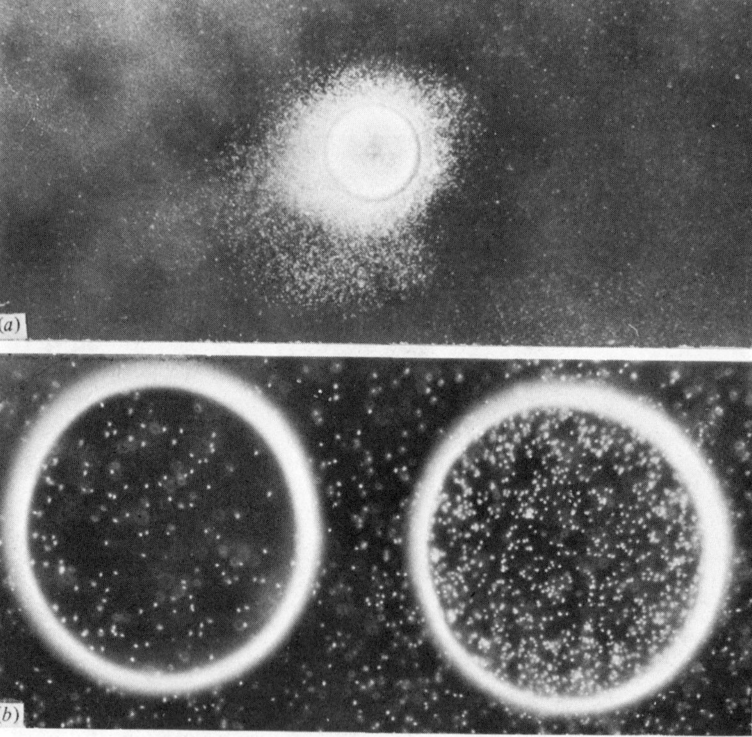

the FC-78 for gametes of *E. siliculosus* are 3.1 μM for ectocarpen, 1 mM for multifiden and *n*-hexane, and for *C. multifida* are 3.1 μM for multifiden, 31 μM for ectocarpen and 1 mM for hexane (Müller, 1976*a*). The true sensitivities of the gametes to the attractants are certainly much higher than these concentrations suggest, as these hydrocarbons have very low solubilities in water.

Müller (1978) has used darkfield illumination and a television video camera to study the swimming behaviour of male gametes of *E. siliculosus* (Table 5). Cells swimming freely in the seawater medium move in straight tracks. They are very sensitive to surfaces, and in close contact with a surface such as a coverslip, they reduce their speed by about a third and swim in wide counterclockwise loops (as viewed from 'inside the medium'). In close proximity to droplets of solvent FC-78 containing ectocarpen they swim more slowly and in tighter circles with an increase in concentration. In the vicinity of a settled female cell, i.e. a point source of ectocarpen, they swim in tight circles or loops, with many sharp turns (Fig. 12). These result either in them swimming in concentric circles around the female, or in them following eccentric paths, reacting strongly when passing the female and so sometimes going astray. However, in these experiments, the female cells were of an American isolate, and the male cells of a European isolate, and these two are incompatible, as no surface–surface recognition takes place (Müller, 1976*b*). In compatible matings, cell–cell contact occurs rapidly between the tip of the male anterior flagellum and the surface of the female cell, resulting in fusion of the gametes.

The cell body of the male gamete is asymmetric, as it is laterally biflagellate, with a leading and a trailing flagellum. This asymmetry must be reflected in the looping and circling swimming behaviour in contact with surfaces.

Table 5. *Swimming characteristics of male gametes of* Ectocarpus siliculosus [a]

Surface contact	Ectocarpen [b]	Swimming pattern	Velocity ($\mu m\ s^{-1}$)	Diameter of circles/loops (μm)	Sharp turns/s^{-1} (angle > 45°)
None	None	Straight	270	—	0
Coverslip	None	Wide loops	190	140	0.2
Solvent	10^{-4}M	Circles	200	100	0.7
Solvent	10^{-3}M	Circles	150	50	2.0
Coverslip	From female cell	Circles/loops	170	50	1.5

[a] Compiled from Müller (1978).

[b] Ectocarpen in droplet of solvent FC-78, or from female gamete under coverslip.

Sex attractants of ciliated protozoa

There is only one example of sexual chemotaxis that has been elucidated in the ciliates. Cells of *Blepharisma* species occur as two mating types, I and II. Type I cells constitutively produce species-specific glycoproteins, blepharmones, which induce type II cells to release, or greatly to increase the production of, blepharismone, the calcium salt of an aromatic acid (Fig. 5; Miyake, 1980). These two hormones regulate the fusion of compatible cells leading to successful mating.

Blepharismone is probably biosynthesized via tryptophan. It induces clumping of type I cells, but also is a potent attractant of type I cells. 5-Hydroxytryptophan inhibits both responses of type I cells to blepharismone, probably acting as an analogue. Thus there could be a common primary site of chemoreception for both responses. Cycloheximide, however, inhibits pairing but not chemotaxis, presumably by its action as an inhibitor of protein synthesis.

Mechanisms of these responses

Receptors

The receptors to these chemical stimuli are most likely to be found in the plasma membrane. In the flagellates and ciliates they could well be localized

Fig. 12. Swimming tracks of 21 male gametes (strain Na-164, Mediterranean) in the vicinity of one female gamete (strain WH-8b, U.S.A.) of *Ectocarpus siliculosus*. Darkfield exposure. Five seconds, field of view is 600×440 μm. (From Müller (1976*b*).)

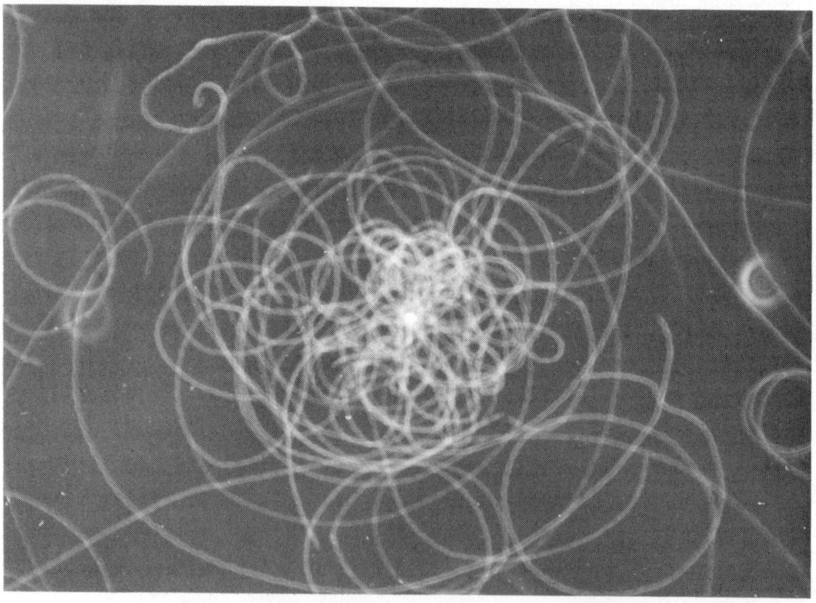

on the membranes enclosing the flagella and cilia themselves, so that these structures would play a sensory role as well as a locomotory one. In the case of stimuli that are ubiquitous metabolites, such as carbohydrates and amino acids, the receptors may be associated with sites of transport, so that the chemosensory system may have evolved from, or together with, an uptake system. In the case of a unique molecule, such as sirenin, it seems to ask a lot for evolution to provide production and reception systems separately and simultaneously, and so perhaps the receptor may have evolved from a membrane-bound enzyme involved in the biosynthesis of sirenin that would already have specificity for binding sirenin.

For bacterial chemosensory transduction, the stimuli from several primary receptors are processed via secondary receptors, and Carlile (1980*a*) suggests that similar transducer proteins may exist in eukaryotic microbial chemoreception.

Control of motility

Currently, the control of motility in the eukaryotic microbes is best understood in *Paramecium* species (Eckert, 1972; Van Houten, 1979). The direction and frequency of beating of the cilia are regulated by the concentration of free intracellular Ca^{++}, and this is regulated by electrical responses of the cell membrane to stimuli. Depolarization, which may be caused by a repellent, leads to an influx of Ca^{++} and a reversal of ciliary beating, and so an avoiding reaction. As the Ca^{++} is pumped out, the cilia resume their normal orientation, and so the cell swims forward again. These Ca^{++} channels are localized in the plasma membrane covering the cilia. Hyperpolarization, which may be caused by an attractant, appears to regulate ciliary beating in a less-direct fashion via the rectifying K^+ conductances associated with the Ca^{++} conductances (Eckert & Brehm, 1979). Hyams & Borisy (1978) using detached flagellar pairs of *Chlamydomonas reinhardii* have demonstrated the same system: membrane depolarization leads to Ca^{++} influx, reversal of flagellar beating, and change in direction of swimming, and then restoration of membrane potential as the Ca^{++} is pumped out leads to forward swimming. For both *Paramecium* and *Chlamydomonas* the critical internal Ca^{++} concentration is about 1 μM; below this level there is 'normal' beating of cilia and flagella, and above it there is 'reversed' beating.

Ca^{++} activities are implicated in several of the other systems discussed here: in *P. polycephalum* oscillations of Ca^{++} concentration occur during streaming, and contraction of the plasmodium occurs in a region of elevated Ca^{++} concentration (Ridgeway & Durham, 1976); Ca^{++} is required for chemotaxis of *Allomyces* gametes (Machlis, 1973) and *Crypthecodinium cohnii* (Hauser, Petrylak, Singer & Levandowksy, 1978).

Future prospects

These microbial systems hold out good hopes for two approaches in the immediate future: the characterization of the individual receptor molecules, and the further elucidation of the bioelectrical control of cell behaviour. From the descriptions of the organisms discussed in this chapter, the challenge clearly is to advance on both fronts in one chosen organism.

I thank the numerous authors who have kindly provided me with reprints and information, and Drs Cameron, Carlile, Dryl, Müller and Pommerville for photographs.

References

Allen, R. N. & Harvey, J. D. (1974). Negative chemotaxis of zoospores of *Phytophthora cinnamomi. Journal for General Microbiology,* **84,** 28–38.

Allen, R. N. & Newhook, F. J. (1973). Chemotaxis of zoospores of *Phytophthora cinnamoni* to ethanol in capillaries of soil pore dimensions. *Transactions of the British Mycological Society,* **61,** 287–302.

Allen, R. N. & Newhook, F. J. (1974). Ethanol suppresses spontaneous turning activity in zoospores of *Phytophthora cinnamomi. Transactions of the British Mycological Society,* **63,** 383–5.

Bean, B. (1979). Chemotaxis in unicellular eukaryotes. In *Physiology of Movements,* ed. W. Haupt & M. E. Feinleib, pp. 335–82. Berlin: Springer-Verlag.

Cameron, J. N. & Carlile, M. J. (1978). Fatty acids, aldehydes and alcohols as attractants for zoospores of *Phytophthora palmivora. Nature, London,* **271,**448–9.

Carlile, M. J. (1975). *Primitive Sensory and Communication Systems.* London: Academic Press.

Carlile, M. J. (1980*a*). Sensory transduction in aneural organisms. In *Photoreception and Sensory Transduction in Aneural Organisms,* ed. F. Lenci & G. Calombetti, pp. 1–22. New York: Plenum Press.

Carlile, M. J. (1980*b*). Positioning mechanisms–the role of motility, taxis and tropism in the life of microorganisms. In *Contemporary Microbial Ecology,* ed. D. C. Ellwood, J. N. Hedger, M. Latham, J. M. Lynch & J. H. Slater, pp. 55–74. London: Academic Press.

Dryl, S. (1963). Contributions to mechanism of chemotactic response in *Paramecium caudatum. Animal Behaviour,* **11,** 393–6.

Dryl, S. (1973). Chemotaxis in ciliate Protozoa. In *Behaviour of Microorganisms,* ed. A. Peŕez-Miravete, pp. 16–30. London: Plenum Press.

Eckert, R. (1972). Bioelectrical control of ciliary activity. *Science, New York,* **176,** 472–81.

Eckert, R. & Brehm, P. (1979). Ionic mechanisms of excitation in *Paramecium. Annual Review of Biophysics and Bioengineering,* **8,** 353–85.

Hauser, D. C. R., Petrylak, D., Singer, G. & Levandowksy, M. (1978). Calcium-dependent sensory-motor response of a marine dinoflagellate to CO_2. *Nature, London,* **273,** 231–2.

Hirschberg, R. & Rodgers, S. (1978). Chemoresponses of *Chlamydomonas reinhardtii. Journal of Bacteriology,* **134,** 671–3.

Hyams, J. S. & Borisy, G. G. (1978). Isolated flagellar apparatus of *Chla-*

mydomonas: characterization of forward swimming and alteration of waveform and reversal of motion by calcium ions *in vitro*. *Journal of Cell Science,* **33,** 235–53.

Khew, K. L. & Zentmeyer, G. A. (1973). Chemotactic response of zoospores of five species of phytophthora. *Phytopathology,* **63,** 1511–17.

Knowles, D. J. C. & Carlile, M. J. (1978). The chemotactic response of plasmodia of the myxomycete *Physarum polycephalum* to sugars and related compounds. *Journal of General Microbiology,* **108,** 17–25.

Levandowsky, M. & Hauser, D. C. R. (1978). Chemosensory responses of swimming algae and protozoa. *International Review of Cytology,* **53,** 145–210.

Machlis, L. (1973). Factors affecting the stability and accuracy of the bioassay for the sperm attractant sirenin. *Plant Physiology, Lancaster,* **52,** 524–6.

Madelin, M. F., Audus, F. & Knowles, D. (1975). Chemotactic and other responses of plasmodia of *Badhamia utricularis* to an extract of *Stereum hirsutum* and to certain other substances. *Journal of General Microbiology,* **89,** 235–44.

Margulis, L. (1970). *Origin of Eukaryotic Cells.* New Haven and London: Yale University Press.

Miyake, A. (1980). Cell interaction by gamones in *Blepharisma.* In *Sexual Interactions in Eukaryotic Microbes,* ed. D. H. O'Day & P. A. Horgen New York: Academic Press, pp. 95–129.

Müller, D. G. (1974). Sexual reproduction and isolation of a sex attractant in *Cutleria multifida* (Smith) Grev. (Phaeophyta). *Biochemie und Physiologie der Pflanzen,* **165,** 212–15.

Müller, D. G. (1976*a*). Quantitative evaluation of sexual chemotaxis in two marine brown algae. *Zeitschrift für Pflanzenphysiologie,* **80,** 120–30.

Müller, D. G. (1976*b*). Sexual isolation between a European and an American population of *Ectocarpus siliculosus* (Phaeophyta). *Journal of Phycology,* **12,** 252–4.

Müller, D. G. (1978). Locomotive responses of male gametes to the species specific sex attractant in *Ectocarpus siliculosus* (Phaeophyta). *Archiv für Protistenkunde,* **120,** 371–7.

Müller, D. G. & Seferiadis, K. (1977). Specificity of sexual chemotaxis in *Fucus serratus* and *Fucus vesiculosus* (Phaeophyceae). *Zeitschrift für Pflanzenphysiologie,* **84,** 85–94.

Orpin, C. G. & Bountiff, L. (1978). Zoospores chemotaxis in the rumen phycomycete *Neocallimastix frontalis.* *Journal of General Microbiology,* **104,** 113–22.

Pommerville, J. (1977). Chemotaxis of *Allomyces* gametes. *Experimental Cell Research,* **109,** 43–57.

Pommerville, J. (1978). Analysis of gamete and zygote motility in *Allomyces.* *Experimental Cell Research,* **113,** 161–72.

Pommerville, J. (1980). The role of sexual pheromones in *Allomyces.* In *Sexual Interactions in Eukaryotic Microbes,* ed. D. H. O'Day & P. A. Horgen New York: Academic Press, pp. 53–72.

Ridgeway, E. B. & Durham, A. C. H. (1976). Oscillations of calcium ion currents in *Physarum polycephalum.* *Journal of Cell Biology,* **69,** 223–6.

Sjoblad, R. D., Chet, I. & Mitchell, R. (1978). Chemoreception in the green algae *Dunaliella tertiolecta.* *Current Microbiology,* **1,** 305–7.

Ueda, T. & Kobatake, Y. (1977). Changes in membrane potential, zeta potential and chemotaxis of *Physarum polycephalum* in response to *n*-alcohols, *n*-aldehydes and *n*-fatty acids. *Cytobiologie,* **16,** 16–26.

Van Houten, J. (1977). A mutant of *Paramecium* defective in chemotaxis. *Science, New York,* **198,** 746–8.

Van Houten, J. (1979). Membrane potential changes during chemokinesis in *Paramecium. Science, New York,* **204,** 1101–3.

Young, B. R., Newhook, F. J. & Allen, R. N. (1979). Motility and chemotactic response of *Phytophthora cinnamomi* zoospores in 'ideal soils'. *Transactions of the British Mycological Society,* **72,** 395–401.

GERALD L. HAZELBAUER

The molecular biology of bacterial chemotaxis

Many species of bacteria are motile as a result of the functioning of bacterial flagella. Even these simple cells possess a sensory system that detects specific chemical compounds and influences motile behaviour so that cells move toward favourable environments. The study of chemotactic behaviour of bacteria was initiated in the nineteenth century by several distinguished botanists and microbiologists, most notably Wilhelm Pfeffer (1904) whose descriptions of chemotaxis included a wide range of unicellular organisms. In recent years, the study of chemotaxis by the enteric bacteria, *Escherichia coli* and *Salmonella typhimurium* initiated by Julius Adler about fifteen years ago, has become an active and growing area of molecular biological research. The exciting possibility is that the strategy of studying a biological phenomenon in the simplest organism in which it occurs will prove as productive in the study of sensory systems as it has been in the elucidation of basic mechanisms of synthesis and control of biological macromolecules. In fact, significant progress has been made toward a molecular and mechanistic description of chemotactic behaviour by enteric bacteria. In this article I will outline our current understanding of the bacterial chemosensory system, with particular emphasis on the role of receptors and the way in which information is transmitted from receptors to effectors.

Motility of flagellated bacteria

Bacteria respond to changes in their chemical environment by altering their pattern of motility. The responses enable the cells to migrate along chemical gradients toward higher concentrations of some compounds (attractants) and away from higher concentrations of others (repellents). *E.coli* and many other bacteria alternate between two types of motility: swimming in which forward progress is made, and tumbling, during which the cell is reoriented to a new, randomly chosen direction (Berg, 1975*a*). Swimming is interrupted every few seconds by episodes of tumbling lasting tenths of seconds: thus cells trace a random walk in three dimensions. In the absence of an orienting gradient there is no net progress in any particular direction. Cells respond to changes in concentration of active compounds over time (temporal gradients) and over distance (spatial gradients) by altering the frequency of tumbling (Berg & Brown, 1972; Macnab & Koshland, 1972). Favourable changes (attractant increases, repellent decreases) inhibit tumbling and unfavourable ones increase tumbling. For cells moving in a spatial gradient, tactic responses bias the random walk by increasing the swimming time in favourable directions, resulting in net migration along the gradient; that is to say, chemotaxis.

The modes of swimming and tumbling can be understood in the context of the biologically unique structure and function of bacterial flagella (see Berg, 1975*b*, for a more extensive discussion). These organelles are totally unlike flagella and cilia of eukaryotic organisms. The complex structures in higher cells consist of a characteristic array of ten sets of tubules surrounded by a membrane. Eukaryotic flagella push cells with a wave motion powered by energy derived from the splitting of adenosine triphosphate, in a manner related to the action of the actomyosin of muscle cells. Bacterial flagella consist of a long filament, a left-handed helix of protein subunits, with a diameter about that of a single eukaryotic tubule and a basal end embedded in the cell membrane. The basal end functions as a rotary motor, turning the filament like a propeller. Rotation is powered by proton motive force, the sum of the electrochemical and hydrogen ion potential across the membrane. Anticlockwise rotation of the left-handed helices results in a concerted pushing force by the several flagella of *E.coli*, producing swimming (Macnab, 1977). Clockwise rotation results in deformation of the filaments and uncoordinated pulling forces, producing tumbling (Macnab & Ornston, 1977). Thus the behavioural pattern of swimming and tumbling reflects the alternation between anticlockwise and clockwise rotation of the flagellar rotary motor.

Adaptation

The bacterial response to a change in the chemical environment is transient. Addition of an attractant to a cell suspension results in an immediate suppression of tumbles as a response to the temporal gradient. Addition of a repellent induces immediate and continual tumbling. However, after a time ranging from seconds to several minutes, depending upon the compound and the magnitude of the gradient, the cells resume their original pattern of swimming and tumbling, even though the attractant or repellent is still present. Thus bacteria, like many sensory cells in higher organisms, adapt to the continued presence of a stimulus. The occurrence of adaptation implies that the sensory system, is functionally sensitive to changes in the chemical environment, rather than to absolute concentrations of active compounds. This feature is logical in relation to the function of the sensory system in directing the cell along gradients. (See also the first chapter of this book for a discussion of the effects of adaptation.)

The transient sensory response can be considered to occur in two stages, excitation and adaptation. Excitation occurs when recognition of active compounds results in a shift in the balance between anticlockwise and clockwise rotation of the flagellar motor. Similar phenomena are often observed in the study of the control of gene or protein activity. For instance, binding of a small molecule at a control site of an allosteric enzyme or regulatory protein will shift an equilibrium between an active and an inactive state of the protein. However, such shifts persist as long as the small molecule is present. It is adaptation that distinguishes a sensory response from these other phenomena. The distinction between chemoresponses with and without adaptation is illustrated in Fig. 1.

Chemoreceptors

Approximately ten years ago Adler (1969), established that chemotaxis in *E. coli* is mediated by specific receptors. In the following years, one-half of

Fig. 1. Contrasting responses of *E. coli* cells to addition of maltose. A temporal gradient from 0 to 1 mM maltose causes an immediate, maximal sensory response (x) in the form of suppression of all tumbles. However, adaptation results in resumption of tumbles at the original frequency in the population, with a half-time of about seven minutes. In contrast, addition of maltose to uninduced cells results in an induction response (dashed line), a gradual increase in the cellular content of enzymes of maltose metabolism (circles) and in the proteins of the maltose transport system (triangles) to a maximal level that is maintained as long as maltose is present. (From Hazelbauer (1980) with permission from *Endeavour*.)

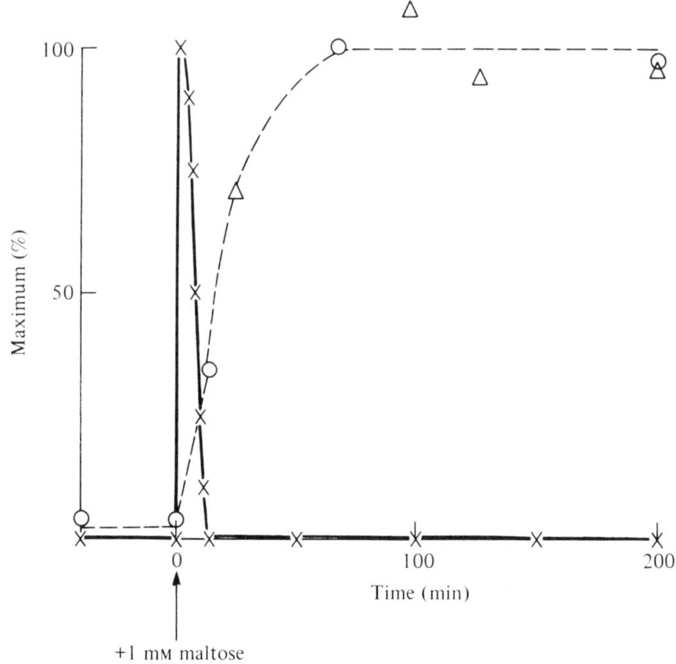

the approximately twenty-five chemoreceptor classes, originally deduced from the results of inhibition experiments, have been identified with specific cell-surface proteins by genetic and biochemical studies. Three receptors – those for galactose-glucose, maltose, and ribose – can be easily obtained as pure proteins and, in fact, were some of the first cell surface receptors to be identified and characterized as pure species (Hazelbauer & Adler, 1971; Hazelbauer & Parkinson, 1977). The affinity and specificity of the binding sites of the three receptors have been determined by studies using the purified proteins. For example, the galactose-glucose receptor binds galactose or glucose with high affinity, fucose with moderate affinity, arabinose poorly, and glycerol not at all. The tactic responses of whole cells to spatial or temporal gradients of these compounds correspond to receptor binding. Galactose and glucose are good attractants; fucose attracts only at relatively high concentrations; there is little response to arabinose; and none at all to glycerol. Binding of ligand to pure receptor can be described by simple mass-action relationships. For the three purified receptors there appears to be a single site per receptor molecule exhibiting a characteristic dissociation constant for the best ligands of approximately 1 μM. Thus there is little tactic response to gradients at concentrations below one-tenth of this value and the response approaches a maximal value at ten- to a hundred-fold above that concentration. Fig. 2 illustrates some of these relationships. As will be seen in the following sections, it appears that receptor proteins function as reporter molecules, sending information about the current concentration of ligand, as reflected by the proportion of occupied binding sites, to a central mechanism that regulates the flagellar rotary motors.

It appears that there are two rather different sorts of chemoreceptors in *E.coli*. First, the better characterized type are the sugar receptors, which perform two separate roles. In addition to being chemoreceptors, the proteins are the substrate-binding components of specific systems that transport the particular ligands across the cytoplasmic membrane (Hazelbauer & Parkinson, 1977). Mutational analysis reveals that the respective tactic and transport systems are otherwise functionally independent. The receptors for galactose-glucose (Hazelbauer & Adler, 1971), maltose (Hazelbauer, 1975) and ribose (Aksamit & Koshland, 1974) are the respective hydrophilic sugar-binding proteins found in the periplasmic space, the compartment between the cytoplasmic and outer membranes of Gram-negative bacteria. All other sugar attractants appear to be recognized by enzymes II (Adler & Epstein, 1974), the sugar-specific, membrane components of the phosphotransferase transport system first described by Kundig & Roseman (1971). Those sugars include glucose, mannose, fructose, *N*-acetylglucosamine, mannitol, sorbitol, and galactitol. From a genetic viewpoint, both the periplasmic binding proteins and the enzymes II are primarily transport proteins and only secondarily chemotactic components. The respective genes occur in operons which also include genes related to transport or metabolism and are induced by their respective transport substrates. In contrast, all other chemotactic components are under the same genetic control as proteins involved in the structure or assembly of flagella (Silverman & Simon, 1977*a*). A reasonable interpretation of these observations is that the sugar receptors are transport-system recognition components that were secondarily recruited as chemoreceptors.

The second type of receptors, those which recognize amino acids, are less well understood. However, in general, amino acids are much more powerful attractants than sugars. Recent studies (Clarke & Koshland, 1979) indicate that the receptors for the two best attractants, aspartate and serine, are cytoplasmic membrane proteins that are under the same control as flagellar components and do not seem to be related to transport systems. Thus it seems that the sensory system in enteric bacteria was designed primarily to respond to amino-acid stimuli and only secondarily to sugar stimuli.

Fig. 2. Correlation of chemotactic response with binding of attractant to receptor. The per cent occupancy of the pure maltose-binding protein with maltose (filled circles) at different concentrations of free sugar is compared to the tactic behaviour of whole cells migrating in spatial gradients (hollow circles) or responding to temporal gradients (triangles) of maltose. (From Hazelbauer (1980) with permission from *Endeavour*.)

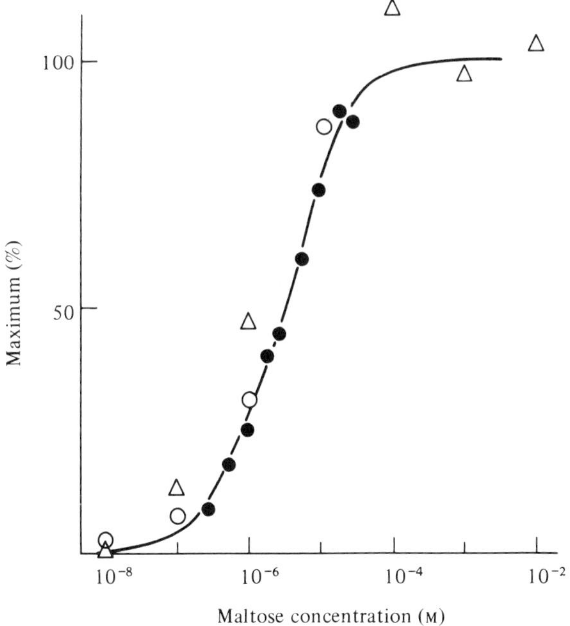

The tumble regulator

The central mechanism is called the tumble regulator, and can be defined by the functions it performs. The regulator accepts information from all receptors, and processes that information to produce a unified signal that controls the direction of rotation of the flagellar motor. Components of the regulator have been defined genetically by mutations that disrupt these central functions and thus produce cells unable to perform normal chemotaxis toward or away from any active compound. Generally non-chemotactic mutants define six separate *che* genes in *E.coli* (Parkinson, 1977). The products of these genes are all water-soluble proteins located inside the cell (Silverman & Simon, 1977*b*). There is genetic and biochemical evidence that some of the *che* gene products interact with membrane components, including proteins of the flagellar motor (Parkinson & Parker, 1979). Some of the genes for flagellar proteins can be mutated in such a way that the flagellar motor still functions in mutant cells but the balance between counterclockwise and clockwise rotation is drastically shifted to one extreme or the other. Thus it seems that some flagellar proteins are functionally part of the tumble regulator.

Transducers

Stimuli recognized by twenty-five different types of receptors must focus on the tumble regulator. Initially the stimulus is the physical presence of more or fewer molecules of an active compound. Active molecules are bound by receptors, and changes induced by the binding are thought to function as informational signals which pass to the tumble regulator. Thus in passing from receptor to tumble regulator, tactic stimuli are transduced into information-carrying signals and transmitted from one physical location to another. The components that link receptors to the tumble regulator are called transducers. Mutations in a gene coding for a transducer eliminate responses mediated by the receptor(s) linked to that transducer but do not affect responses mediated by receptors linked to other transducers. Three classes of transducer mutations – *tsr, tar,* and *trg* – have been identified and each class eliminates responses mediated by a special group of receptors (Springer, Goy & Adler, 1977; Hazelbauer & Harayama, 1979). Thus it appears that transducers serve to focus signals from several receptors. None of the three known types of transducer mutations eliminate responses to enzyme II sugars, so there is likely to be at least a fourth transducer linked to enzyme II receptors (Hazelbauer & Engstroṁ, 1980). Studies of the maltose receptor have been shown that there is a direct interaction of ligand-occupied receptor with the relevant transducer protein (the *tar* product) (Koiwai & Hayashi, 1979). A number of lines of evidence suggest that known transducer proteins all interact directly with their respective families of receptors and transmit the resulting signals directly to components of the tumble regulator. A diagram of the flow of tactic information in *E. coli* is shown in Fig. 3.

It is interesting that grouping of receptors in linkage to a common transducer parallels the division into two classes of receptors. Amino-acid receptors are linked to *tsr* or *tar,* while sugar receptors are linked to *trg* or the postulated enzyme II-related transducer. Responses to repellents and the receptors that mediate these responses (Tso & Adler, 1974) are not well understood, so it is not possible to classify repellent responses as weak or strong in the same terms as one can classify responses to attractants as weak (sugars) or strong (amino acids). In any case, apparently all of the nine receptor classes defined for repellents are linked to the amino acid-related transducers. There is one exceptional sugar receptor, for maltose, that is linked to *tar,* a transducer otherwise linked to amino-acid receptors. Maltose stimuli are six times more effective than galactose or ribose stimuli even when normalized to the number of receptors (Koman, Harayama & Hazelbauer, 1979). This observation suggests that the greater effectiveness of amino acids as attractants may be a function of the transducers to which the receptors are linked. The difference among transducers may simply reflect the number of molecules per cell. There are 1000–3000 tsr and tar proteins per cell but probably only several hundred trg proteins (Hazelbauer & Engström, 1981).

Protein carboxyl methylation

Methionine is required for chemotaxis. In *E.coli,* mutants unable to synthesize this amino acid are able to migrate in spatial gradients only if an

Fig. 3. Flow of tactic information in the *E.coli* sensory system. Repellent receptors are written in bold face type. (From Hazelbauer (1980) with permission from *Endeavour*.)

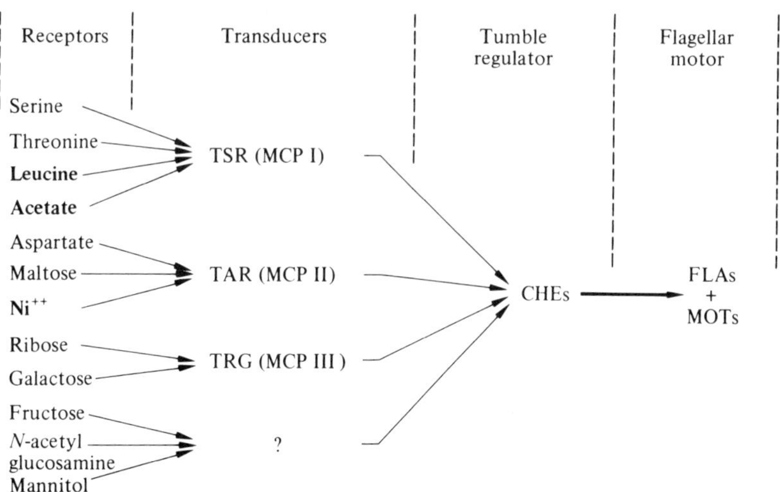

exogenous source of methionine is available. In 1975, in Adler's laboratory, a specific biochemical reaction was identified that involves a derivative of methionine and is intimately involved in chemotactic behaviour (Kort, Goy, Larsen & Adler, 1975). Specifically, a group of proteins located in the cytoplasmic membrane are covalently modified by donation of methyl groups from *S*-adenosyl methionine, an activated form of methionine, to create carboxyl methyl esters of certain glutamyl residues in the proteins (see Fig. 4). In the past several years, work from the laboratories of Adler, M. Silverman & M. Simon and D. E. Koshland, Jr. has done much to define the details and significance of this reaction (Hazelbauer, 1979; Springer, Goy & Adler, 1979). Before discussing protein methylation, it would be appropriate to consider some of the experimental procedures by which it is studied.

The methylation of proteins can be demonstrated by providing cells with methionine containing a radioactive atom in the donatable methyl group. When protein synthesis is inhibited by an appropriate antibiotic, then proteins in the

Fig. 4. The methylation cycle of methyl-accepting chemotaxis proteins (MCPs). (From Hazelbauer (1980) with permission from *Endeavour*.)

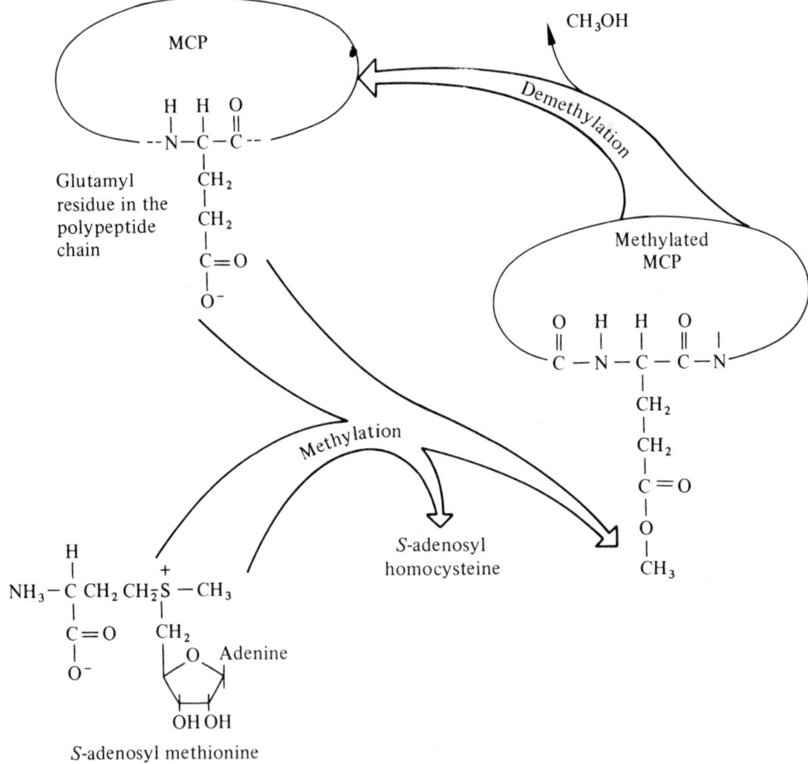

cell can acquire the radioactive label only by being methylated. The proteins can be sorted into a linear array of bands, according to their molecular weights, by electrophoresis in polyacrylamide gels in the presence of the strong, ionic detergent, sodium dodecyl sulphate. After electrophoresis, a gel can be dried on to filter paper and subsequently the pattern of protein bands can be analysed to locate radioactivity (that is, labelled methyl groups) by autoradiography or fluorography. Examples of such fluorograms showing the methyl-accepting chemotaxis proteins (MCPs) can be seen in Fig. 5.

In unstimulated cells there is a basal level of methylation. This level changes during adaptation of cells to temporal gradients. Adaptation to favourable gradients corresponds to increases in methylation while adaptation to unfavourable gradients corresponds to decreases. The time courses of adaptation and of changes in methylation are coincident. Extensive studies by M. Springer and M. Goy in Adler's laboratory have provided convincing evidence that protein carboxyl methylation is intimately related to the process of adaptation and probably represents the biochemical mechanism of that important function. Goy & Springer (1978) summarize those studies.

Methylation is catalysed by a specific methyltransferase (Springer & Kosh-

Fig. 5. Patterns of methylation of MCPs of wild-type (w.t.) and mutant bacteria. A fluorogram of the region of a sodium dodecyl sulphate polyacrylamide gel showing [*Me*-^{3}H]-labelled methyl-accepting chemotaxis proteins. Wild-type cells (lanes 1–4), *tsr* mutant (MCP I) cells (lanes 7 and 8), and *tar* mutant (MCP II) cells (lanes 5 and 6) in the presence of [*Me*-^{3}H]methionine were left unstimulated (lanes 1,5,7), or were allowed to adapt to stimuli of 10 mM serine (lane 2), 10 mM α-methylaspartate (αma) a non-metabolizable analogue of aspartate (lanes 3,8), or 50 mM α-aminoisobutyric acid (aibu) a nonmetabolizable analogue of serine (lanes 4,6). Note that serine is recognized by both a high-affinity receptor linked to MCP I and a low-affinity receptor linked to MCP II, while aibu is recognized only by the high-affinity receptor. (From Engstroñ & Hazelbauer (1980) with permission from MIT Press.)

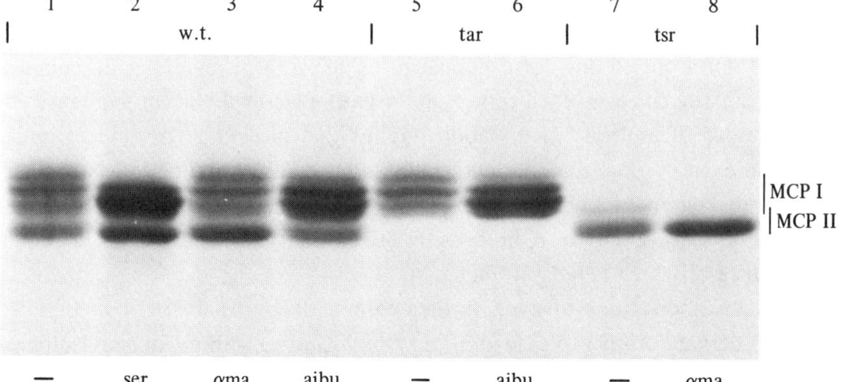

land, 1977), and demethylation by a specific demethylase (Stock & Koshland, 1978). The enzymes are products of two of the *che* genes, *che*R and *che*B, respectively. In the absence of either enzyme activity, cells are unable to perform normal chemotaxis.

Three different MCPs are known, all with similar molecular weights of about 60 000. MCP I and MCP II are the products of the transducer genes *tsr* and *tar*, respectively (Springer, Goy & Adler, 1977; Silverman & Simon, 1977*a*), and MCP III is the product of *trg* (Kondoh, Ball & Adler, 1979; Hazelbauer, Engström & Harayama, 1981). Adaptation to tactic stimuli transduced through a specific transducer is achieved by an increase or decrease in methylation of that particular transducer protein. Thus the transducer MCPs are central to both excitation as transducers of stimuli and to adaptation as acceptors in methylation. The two functions can be separated. Cells unable to methylate owing to lack of methionine or of the methyl transferase still respond to appropriate stimuli but are unable to adapt to them (Goy, Springer & Adler, 1978; Parkinson & Revello, 1978).

The MCPs are clearly complex proteins both functionally and structurally. Since MCPs interact with receptors found on the outer side of the cell membrane and with methylating and demethylating enzymes located inside the cell, the proteins must span the membrane and have sites for the relevant interactions both inside and on the outside of the cell. Recent work carried out independently in several laboratories provides evidence for further complexity (Engström & Hazelbauer, 1980; Chelsky & Dahlquist, 1980; De Franco & Koshland, 1980; Boyd & Simon, 1980). The conclusion reached by all these workers is that each MCP molecule has more than one glutamyl residue that can be methylated. Addition of each methyl group causes a slight downward shift in the position of the molecule on the polyacrylamide gel. Thus an individual MCP is seen in the gel pattern as a set of bands (Fig. 5), reflecting different numbers of methyl groups per polypeptide chain. There are probably three or four methylation sites on MCP I or MCP II and two on MCP III. It appears that all sites can be methylated upon adaptation to any particular stimulus (Hazelbauer *et al.*, 1981), so there is probably not a one-to-one correspondence on the transducer molecule between the binding site for a given receptor and a particular methylation site. Instead, the existence of multiple methylation implies that individual transducer molecules can exist in more than the two states suggested by simple models of alternation between an excited and an adapted (methylated) state (Fig. 6). A transducer/MCP may exist in a graded series of adapted states, the functional significance of which is not yet understood.

There are indications of even further complexities. Analysis of methylated peptides derived from protease digestion of different sodium dodecylsulphate (SDS) gel bands of a given MCP indicate that there is a non-random order of methylation of the various accepting sites (Engström & Hazelbauer, 1980; Chelsky & Dahlquist, 1980). Two-dimensional gels, separating components by isoelectric focusing in one dimension and electrophoresis in the presence of SDS in the others, resolve MCP II into a pattern of six spots (Fig. 7) and MCP I into a pattern of more than ten spots. These complex patterns are due in part to multiple methylation but it appears that there is at least one additional factor

Fig. 6. Cycles of excitation and adaptation. (Adapted from Parkinson (1977). From Hazelbauer (1980) with permission from *Endeavour*.)

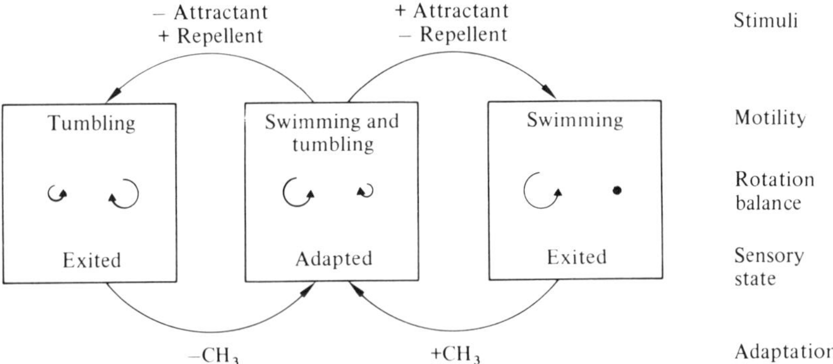

Fig. 7. Multiple forms of MCP II resolved on a two-dimensional gel. Membranes from a wild-type cell labelled with [*Me*-^{3}H] were subjected to isoelectric focusing (higher to lower pH, left to right) in the first dimension and then sodium dodecyl sulphate polyacrylamide (10%) gel electrophoresis in the second dimension (higher to lower apparent mol. wt, top to bottom). The figure is a fluorogram of the region of the gel including MCP II, which appears as two lines of spots distributed along diagonals from acidic, higher apparent mol. wt towards basic, lower apparent molecular weight. The spots are distributed in the pH range 5.8 to 5.9 and an apparent mol. wt range of 60 000 to 62 000.

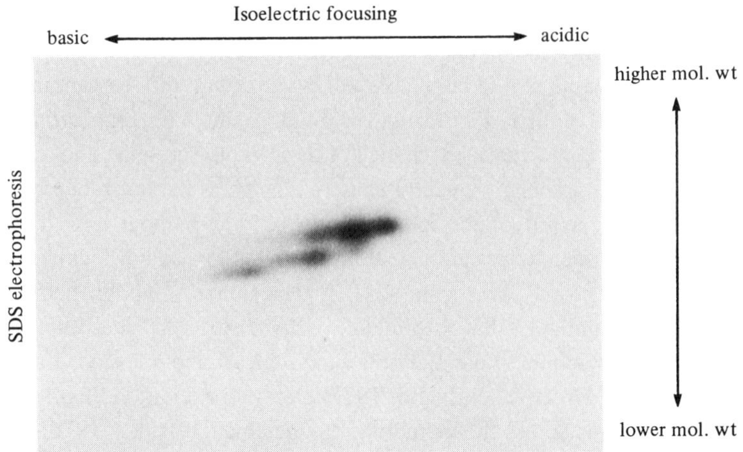

involved in generating the multiplicity of forms (Hazelbauer & Engström, 1981). The functional significance of the additional factor is unknown.

Protein carboxyl methylation in other organisms

Protein carboxyl methylation is not limited to bacteria. Appropriate methyltransferase activity is found in many mammalian tissues and is particularly high in endocrine organs and certain parts of the central nervous system. Specific methyl-accepting proteins have not yet been identified in these tissues, but significant methyl-accepting activity implies that such proteins are probably present. Studies of release of hormones from the adrenal medulla (Diliberto, Viveros & Axelrod, 1976) and the chemotactic behaviour of white blood cells (O'Dea *et al.*, 1978) provide indications that protein carboxyl methylation is involved in the stimulus-response phenomena observed in these systems. It may very well be that this reaction is of general importance in mechanisms linking recognition at receptor sites to response of an effector and thus information obtained in the study of the bacterial system may have wider significance.

How does the chemosensory system work?

The bacterial sensory system must involve at least two processes that are activated by chemical stimulation. The first, excitation, is rapid, occurring in less than a second. The second, adaptation, is relatively slow, occurring over seconds or minutes. Any suggestions about the mechanistic basis of tactic behaviour must thus incorporate a rapid change, excitation, and a slower, balancing alteration, adaptation. In this section, I will summarize what is known about the components involved in the two phases of chemotaxis and suggest ways in which they might function.

Excitation requires an appropriate receptor and transducer and passage of an excitatory signal from the transducer to affect the gearshift that determines the direction of rotation of the flagellar motor. The simplest linkage between receptor and transducer is the one suggested in an earlier section. An attractant receptor without ligand would be unable to associate with the relevant transducer, but upon binding of ligand a transducer-binding site with high affinity would be exposed by a conformational change of the receptor protein. The resulting receptor-transducer association would cause a change in the transducer protein which affects the gearshift of the motor. Little is known about how transducers are linked to motors. It appears that transducer/MCPs are not localized around flagellar basal ends (P. Engström & G. L. Hazelbauer, unpublished observations) and so it is unlikely that the linkage is by direct, physical interaction. Gross changes in membrane potential on a time scale of about a second do not seem to be involved (Miller & Koshland, 1977) although excitation might involve fluxes of specific ions across the membrane (Szmelcman & Adler, 1976).

It is an attractive possibility to consider that a soluble component is in association/dissociation equilibrium with both transducers and motors, and clockwise or anticlockwise rotation is determined by whether the component is off or on the motor. Then a change in affinity of transducer for the component would shift the equilibrium, inducing one or the other directions of rotation, and methylation or demethylation of the transducer could re-establish the initial transducer affinity and thus the initial equilibrium (Parkinson, 1977; Hazelbauer, 1980). If such a soluble excitation component exists, it is not the product of any of the known *che* genes, since none of those genes are required absolutely for excitation (Parkinson, 1977, and personal communication). Ordal (1977) reported that perturbing the level of Ca^{++} ion in *Bacillus subtilis* affects the balance between swimming and tumbling; therefore this ion might be involved in controlling the flagellar motor. The widespread participation of the calcium ion in receptor-mediated phenomena makes its involvement in the bacterial sensory system a particularly interesting possibility.

As indicated previously, the simplest interpretation of the available data is that carboxyl methylation or demethylation of the transducer through which a stimulus passed is the mechanism of adaptation. Whatever the change or signal generated by receptor-transducer interaction, it is functionally cancelled by the appropriate change in methylation level. It is not known whether methylation changes are induced directly by receptor–transducer interactions or indirectly by feedback from flagellar motors that have received excitatory signals. The latter possibility seems somewhat more likely since normal adaptation requires functional products of all six *che* genes and some of those proteins are known to interact with flagellar basal ends (Parkinson & Parker, 1979). Normal adaptation must reflect a complicated balance between the enzymes of methylation and demethylation and their substrates, the MCPs. Elimination of approximately half of the total MCPs in a *tsr* mutant (MCP I^-) results in longer adaptation times. Mutants in *tar* (MCP II^-) show a similar, but less pronounced phenotype. Double *tsr tar* mutants, missing approximately ninety per cent of MCPs, have an extremely low tumble frequency unless left completely undisturbed for extended periods of time, after which normal behaviour ensues. If mutant cells are then stimulated, excitation occurs normally but adaptation is vastly delayed (Hazelbauer & Engström, 1980).

Conclusions

At present we have a reasonable idea of the molecules involved in the bacterial chemotactic system and some notion of the roles they perform. With the elucidation of the importance of protein carboxyl methylation, a first step has been taken toward a description of the biochemical mechanisms of the sensory system. It seems likely that continued application of the combination of genetic and biochemical approaches will produce a detailed understanding of the molecular biology of bacterial chemotaxis. There is even reason for optimism that some of the principles of receptor–effector linkage elucidated for bacteria will be significant for a wide range of receptor-linked phenomena.

References

Adler, J. (1969). Chemoreceptors in bacteria. *Science, New York,* **166,** 1588–97.

Adler, J. & Epstein, W. (1974). Phosphotransferase-system enzymes as chemoreceptors for certain sugars in *Escherichia coli* chemotaxis. *Proceedings of the National Academy of Sciences of the U.S.A.,* **71,** 2895–9.

Aksamit, R. R. & Koshland, D. E., Jr. (1974). Identification of the ribose-binding protein as the receptor for ribose chemotaxis in *Salmonella typhimurium. Biochemistry, New York,* **13,** 4473–8.

Berg, H. C. (1975*a*). How bacteria swim. *Scientific American,* **233,** 36–44.

Berg, H. C. (1975*b*) Chemotaxis in bacteria. *Annual Review of Biophysics and Bioengineering,* **4,** 119–36.

Berg, H. C. & Brown, D. A. (1972). Chemotaxis in *Escherichia coli* analyzed by three-dimensional tracking. *Nature, London,* **239,** 500–4.

Boyd, A. & Simon, M. (1980). Multiple electrophoretic forms of methyl-accepting chemotaxis proteins generated by stimulus-elicited methylation in *Escherichia coli. Journal of Bacteriology,* **143,** 809–15.

Chelsky, D. & Dahlquist, F. W. (1980). Structural studies of methyl-accepting chemotaxis proteins of *Escherichia coli:* evidence for multiple methylation sites. *Proceedings of the National Academy of Sciences of the U.S.A.,* **77,** 2434–8.

Clarke, S. & Koshland, D. E., Jr. (1979). Membrane receptors for aspartate and serine in bacterial chemotaxis. *Journal of Biological Chemistry,* **254,** 9695–702.

De Franco, A. L. & Koshland, D. E., Jr. (1980). Multiple methylation in the processing of sensory signals during bacterial chemotaxis. *Proceedings of the National Academy of Sciences of the U.S.A.,* **77,** 2429–33.

Diliberto, E. J., Viveros, O. H. & Axelrod, J. (1976). Subcellular distribution of protein carboxymethylase and its endogenous substrates in the adrenal medulla: possible role in excitation–secretion coupling. *Proceedings of the National Academy of Sciences of the U.S.A.,* **73,** 4050–4.

Engström, P. & Hazelbauer, G. L. (1980). Multiple methylation of methyl-accepting chemotaxis proteins during adaptation of *E. coli* to chemical stimuli. *Cell,* **20,** 165–71.

Goy, M. F. & Springer, M. S. (1978). In search of the linkage between receptor and response: the role of a protein methylation reaction in bacterial chemotaxis. In *Taxis and Behavior,* ed. G. L. Hazelbauer, pp. 3–34. London: Chapman and Hall.

Goy, M. F., Springer, M. S. & Adler, J. (1978). Failure of sensory adaptation in bacterial mutants that are defective in a protein methylation reaction. *Cell,* **15,** 1231–40.

Hazelbauer, G. L. (1975). The maltose chemoreceptor of *Escherichia coli. Journal of Bacteriology,* **122,** 206–14.

Hazelbauer, G. L. (1979). Bacterial chemotaxis and protein carboxymethylation. *Nature, London,* **279,** 18–19.

Hazelbauer, G. L. (1980). Bacterial chemotaxis: molecular biology of a sensory system. *Endeavour,* **4,** 67–74.

Hazelbauer, G. L. & Adler, J. (1971). Role of the galactose-binding protein in chemotaxis of *Escherichia coli* toward galactose. *Nature, New Biology,* **230,** 101–4.

Hazelbauer, G. L. & Engström, P. (1980). Parallel pathways for transduction of chemotactic signals in *Escherichia coli. Nature, London,* **283,** 98–100.

Hazelbauer, G. L. & Engström, P. (1981). Multiple forms of methyl-accepting chemotaxis proteins distinguished by a factor in addition to multiple methylation. *Journal of Bacteriology,* **145,** 35–42.

Hazelbauer, G. L., Engström, P. & Harayama, S. (1981). Methyl-accepting chemotaxis protein III and the transducer gene *trg. Journal of Bacteriology,* **145,** 43–9.

Hazelbauer, G. L. & Harayama, S. (1979). Mutants in transmission of chemotactic signals from two independent receptors of *E. coli. Cell,* **16,** 617–25.

Hazelbauer, G. L. & Parkinson, J. S. (1977). Bacterial chemotaxis. In *Microbial Interactions,* ed. J. L. Reissig, pp. 59–98. London: Chapman & Hall.

Koiwai, O. & Hayashi, H. (1979). Studies on bacterial chemotaxis IV. Interaction of maltose receptor with a membrane-bound chemosensing component. *Journal of Biochemistry,* **86,** 27–34.

Koman, A., Harayama, S. & Hazelbauer, G. L. (1979). Relation of chemotactic response to the amount of receptor: evidence for different efficiencies of signal transduction. *Journal of Bacteriology,* **138,** 739–47.

Kondoh, H., Ball, C. B. & Adler, J. (1979). Identification of a methyl-accepting chemotaxis protein for the ribose and galactose receptors of *Escherichia coli. Proceedings of the National Academy of Sciences of the U.S.A.,* **76,** 260–4.

Kort, E. N., Goy, M. F., Larsen, S. H. & Adler, J. (1975). Methylation of a membrane protein involved in bacterial chemotaxis. *Proceedings of the National Academy of Sciences of the U.S.A.,* **72,** 3939–43.

Kundig, W. & Roseman, S. (1971). Sugar transport II. Characterization of constitutive membrane-bound enzymes II of the *Escherichia coli* phosphotransferase system. *Journal of Biological Chemistry,* **246,** 1407–18.

Macnab, R. M. (1977). Bacterial flagella rotating in bundles: a study in helical geometry. *Proceedings of the National Academy of Sciences of the U.S.A.,* **74,** 221–5.

Macnab, R. M. & Koshland, D. E., Jr. (1972). The gradient-sensing mechanism in bacterial chemotaxis. *Proceedings of the National Academy of Sciences of the U.S.A.,* **69,** 2509–12.

Macnab, R. M. & Ornston, M. K. (1977). Normal-to-curly flagellar transitions and their role in bacterial tumbling. Stabilization of an alternative quaternary structure by mechanical force. *Journal of Molecular Biology,* **112,** 1–30.

Miller, J. B. & Koshland, D. E., Jr. (1977). Sensory electrophysiology of bacteria: relationship of the membrane potential to motility and chemotaxis in *Bacillus subtilis. Proceedings of the National Academy of Sciences of the U.S.A.,* **74,** 4752–6.

O'Dea, R. F., Viveros, O. H., Axelrod, J., Aswanikumar, S., Schiffman, E. & Corcoran, B. A. (1978). Rapid stimulation of protein carboxymethylation in leukocytes by a chemotactic peptide. *Nature, London,* **292,** 462–4.

Ordal, G. W. (1977). Calcium ion regulates chemotactic behavior in bacteria. *Nature, London,* **270,** 66–7.

Parkinson, J. S. (1977). Behavioral genetics in bacteria. *Annual Review of Genetics,* **11,** 397–414.

Parkinson, J. S. & Parker, S. R. (1979). Interaction of the *che*C and *che*Z gene products is required for chemotactic behavior in *Escherichia coli. Proceedings of the National Academy of Sciences of the U.S.A.,* **76,** 2390–4.

Parkinson, J. S. & Revello, P. T. (1978). Sensory adaptation mutants of *E. coli. Cell,* **15,** 1221–30.

Pfeffer, W. (1904). *Pflanzenphysiologie, vol 2 Kraftwechsel.* Leipzig: Wilhelm Engelmann.

Silverman, M. & Simon, M. (1977*a*). Chemotaxis in *Escherichia coli:* methylation of *che* gene products. *Proceedings of the National Academy of Sciences of the U.S.A.,* **74,** 3317–21.

Silverman, M. & Simon, M. (1977*b*). Identification of polypeptides necessary for chemotaxis in *Escherichia coli. Journal of Bacteriology,* **130,** 1317–25.

Springer, M. S., Goy, M. F. & Adler, J. (1977). Sensory transduction in *Escherichia coli:* two complementary pathways of information processing that involve methylated proteins. *Proceedings of the National Academy of Sciences of the U.S.A.,* **74,** 3312–16.

Springer, M. S., Goy, M. F.& Adler, J. (1979). Protein methylation in behavioral control mechanisms and in signal transduction. *Nature, London,* **280,** 279–84.

Springer, W. R. & Koshland, D. E, Jr. (1977). Identification of a protein methyltransferase as the *che*R gene product in the bacterial sensing system. *Proceedings of the National Academy of Sciences of the U.S.A.,* **74,** 533–7.

Stock, J. B. & Koshland, D. E., Jr. (1978). A protein methylesterase involved in bacterial sensing. *Proceedings of the National Academy of Sciences of the U.S.A.,* **75,** 3659–63.

Szmelcman, S. & Adler, J. (1976). Change in membrane potential during bacterial chemotaxis. *Proceedings of the National Academy of Sciences of the U.S.A.,* **73,** 4387–91.

Tso, W. W. & Adler, J. (1974). Negative chemotaxis in *Escherichia coli. Journal of Bacteriology,* **118,** 560–76.

COLIN H.CLARKE

Motility and fruiting in the bacterium *Myxococcus xanthus*

Introduction

The Myxobacteria are aerobic, frequently pigmented, Gram-negative, flexible, rod-shaped bacteria. They show the properties of gliding motility on surfaces, and of aggregation and cooperative morphogenesis to give fruiting bodies (Fig. 1). The gliding motility, like that shown by some other bacterial groups, does not involve bacterial flagella but is accompanied by the deposition of slime trails. The Myxobacteria can thus, very superficially, be regarded as prokaryotic parallels of the eukaryotic cellular slime moulds, which are discussed by Newell in the fifth chapter of this book.

In addition to their cooperative morphogenesis to give fruiting bodies, some species of Myxobacteria also show cellular morphogenesis. Within fruiting bodies, or on treatment with inducing chemicals, vegetative cells are converted to

Fig. 1. Life cycle of *M.xanthus*.

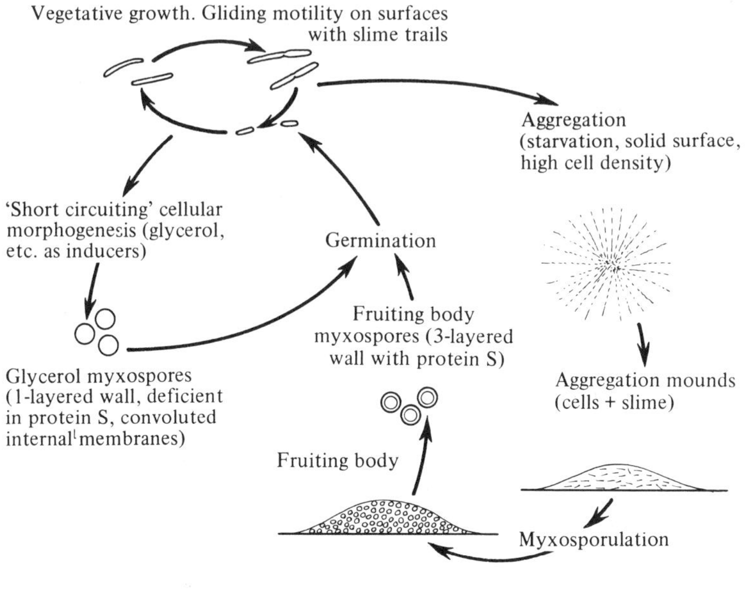

resting cells or to ovoid or spherical thick-walled myxospores (also called microcysts in the older literature). The fruiting bodies in some species are simple tumulus-like mounds of pigmented dried slime containing large numbers of myxospores. In other species the fruiting bodies are of complex structure and may be tree-like in appearance with chambered macrocysts containing large numbers of resting cells (Brockman & Todd, 1974).

The Myxobacteria are found as soil organisms, on rotting vegetation, deposited animal dung, and the bark of trees. In all these environments they are probably acting as micropredators, actively digesting and living off other microorganisms such as bacteria, algae, filamentous fungi and yeasts. Myxobacteria release an impressive battery of extracellular enzymes. A few species are degraders of cellulose, but many will digest starch, glycogen, proteins, and the mucopeptide (murein) layer of both Gram-positive and other Gram-negative bacteria. A high local concentration of such extracellular enzymes results from the tendency of vegetative myxobacterial cells to stay together as moving swarms, behaviour which Dworkin, 1973*a*, has described as that of a 'wolf pack'.

When first isolated from nature most strains of Myxobacteria do not show dispersed growth in liquid media, but in some cases dispersed growth variants, which may or may not retain the ability to form fruiting bodies, can be isolated on repeated subculture. With the exception of some of the cellulolytic species Myxobacteria have quite complex growth-factor requirements. One *Myxococcus xanthus* strain, for example, can be grown, albeit at an extremely poor rate, on a synthetic medium containing pyruvate and aspartate together with the required amino acids isoleucine, leucine and valine, plus either methionine or vitamin B_{12} and phenylalanine, asparagine, spermidine and inorganic salts (Bretscher & Kaiser, 1978). Other amino acids are stimulatory to growth and routinely *M. xanthus* can be grown on a peptide-rich enzymatic digest of casein such as Difco Bacto-Casitone. On casein itself the growth rate has been shown to be dependent upon cell density, due to the 'wolf pack' effect (Rosenberg, Keller & Dworkin, 1977). There is a requirement also for a high concentration of Mg^{++} or Ca^{++} ions. *M. xanthus* can also be grown on living cells of, for example, *Escherichia coli, Micrococcus luteus* or *Saccharomyces cerevisiae*.

The best studied species among the Myxobacteria is undoubtedly *M. xanthus* whose genome has been shown to have a molecular weight of $8.4 \pm 1.2 \times 10^9$ Daltons, some 3.5 times that of the *E. coli* genome (Zusman, Krotoski & Cumsky, 1978). *M. xanthus* forms the simplest, mound, type of fruiting bodies, after cellular aggregation and autolysis under starvation conditions (Fig. 2). Less thoroughly studied have been *Stigmatella aurantiaca* and *Chondromyces apiculatus,* two species whose fruiting bodies are complex, microtree-like, structures. The biology of the myxobacteria has been surveyed extensively in a number of reviews including: Dworkin, 1966, 1972, 1973*a,b*; Sudo & Dworkin, 1973; Wireman & Dworkin, 1975; and most recently, Parish, 1979; and Kaiser, Manoil & Dworkin, 1979.

In *M. xanthus* those aspects of the biology of the bacterium which have been investigated have included 'short circuiting' cellular morphogenesis in which, under the influence of chemical inducers, vegetative cells in liquid growth medium convert synchronously into free myxospores. Glycerol is the archetypal inducer used in such experiments. A number of other aspects of the biology of *M. xanthus* have been studied. These include: the bacteriophages, virulent, temperate and transducing, of this bacterium; conditions governing germination of myxospores; the genetics of the control of gliding motility; the controls of fruiting body formation; and, to lesser degrees, the photobiology of the organism; its ability to act as a host for plasmids transferred into it from enteric bacteria; and a curious phenomenon of yellow-tan colonial colour-phase variation which seems to be connected with fruiting body formation (Wireman & Dworkin, 1975).

In the remainder of this review, I intend to concentrate on the recent analyses of the control of gliding motility and of fruiting body formation. These are not, as will be seen, entirely unconnected topics.

Fig. 2. Fruiting bodies of *M. xanthus* strain MD-1 on water-magnesium-phosphate-agar plus 0.05% (w/v) Oxoid 'Tryptone'. The diameter of the droplet area within which the fruits have formed is approximately 1.5 cm.

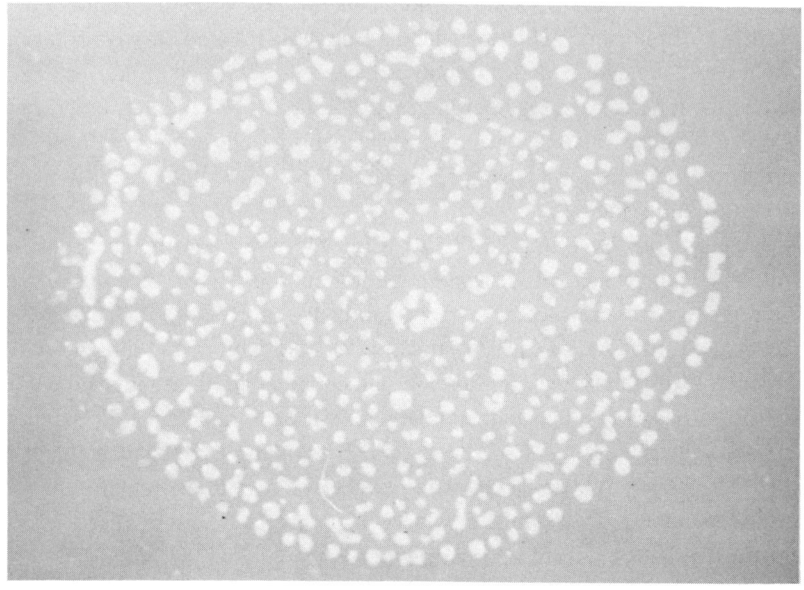

Gliding motility

Gliding motility is expressed by *M. xanthus* both as the outward expansion of cells and swarms at the spreading periphery of the low flat colonies formed on growth medium, and under the very different conditions of aggregation preceding fruiting body formation after starvation. Although under conditions of growth on a rich medium, such as on 2% Casitone–magnesium–agar, all cells may be genotypically motile there is a phenotypic constraint exerted on the motility of those cells in the population which are in the so-called non-swarmer colonial phase. In contrast, swarmer phase cells exhibit good outward gliding motility. Thus with a wild-type population one obtains, on plating out cells on 2% Casitone–magnesium–agar, two main types of colonies. Yellow colonies have, mostly, a so-called 'swarmer' appearance of irregular spreading margin, whereas the tan colonies are mostly of smaller diameter with a much more even margin i.e. 'non-swarmers' (Wireman & Dworkin, 1975). On media with lower concentrations of nutrients both swarmers and non-swarmer cells form colonies with spreading margins. The basis of the swarmer versus non-swarmer control of the phenotypic expression of motility, and its relationship to differences between yellow and tan phase variants is completely unknown, though open to several different speculative interpretations.

Our present understanding of the genetic control, if not as yet the actual mechanism, of gliding motility in *M. xanthus* stems from the observation of yellow-tan phase variation by Burchard & Dworkin (1966). A stable tan colonial colour mutant was isolated from wild-type strain FB and was shown to have a lower rate of colony expansion on solid medium than a yellow colour mutant (Burchard, 1970). From this stable tan mutant Burchard isolated, firstly, a further mutant which he called SM (semi-motile), the cells of which would not glide individually but only when together in groups. From the SM mutant, in a further step, was derived a totally non-motile strain whose cells would glide neither alone nor in groups. Not surprisingly this non-motile derivative was also unable to construct fruiting bodies.

Subsequent work by Burchard (1974) using the stable tan mutant, revealed several interesting aspects of *M. xanthus* gliding motility. Treatment of the cells with the proteolytic enzymes pronase or subtilisin, or 10^{-3} M ethylenediamine tetraacetic acid (EDTA), or osmotic shock, reversibly inhibited gliding motility. This property was recovered after an hour's incubation at 30°C, but such regeneration of gliding was prevented by treatment with chloramphenicol. Burchard (1974) showed that the osmotic shock, which temporarily abolished motility, released into the medium some eighteen periplasmic membrane proteins, as visualized on sodium dodecylsulphate polyacrylamide gels. Antisera raised against these osmotic-shock-released periplasmic proteins inhibited the gliding motility of non-shocked cells causing, perhaps significantly in view of later work (Dobson, McCurdy & MacRae, 1979) on polar pili (fimbriae), pole-to-pole agglutination of cells. Adsorption of the antisera with tan mutant cells fully removed motility-inhibiting antibodies, whereas adsorption with cells of Burchards's semi-motile or totally non-motile mutants only partly removed motility-inhibitory antibodies.

A further, more detailed, immunological analysis of gliding motility is obviously called for in *M. xanthus,* using the various types of motility mutants isolated by Hodgkin & Kaiser (1977). These authors have carried out a most impressive phenotypic and genotypic analysis of gliding motility, which has revealed unexpected complexities in the organization of this process. Firstly there are two, largely separate, sets of genes for gliding motility in *M. xanthus.* Only at a single genetic locus, named *mgl,* do mutations occur which inactivate *both* types of motility which operate in wild-type cells (Kaiser, 1978). These two types of gliding motility are system A (adventurous) which applies to the motility of isolated single cells, and system S (social) which applies to the movements of groups of cells. Thus the semi-motile, SM, mutant isolated earlier by Burchard (1970) is deficient in the A, but proficient in the S, gliding motility system.

By the good fortune which seems to favour the talented, Hodgkin & Kaiser (1977) began their genetic analysis of gliding motility in *M. xanthus* with an, initially unsuspected, spontaneous mutant strain which was A^+ but S^-, i.e. had operative only the motility system for single, but not groups of, cells. Thus from this starting strain they were able to isolate single-step mutants in the A motility system, uncomplicated by the occurrence of social (S) gliding. Genetic analyses, by transduction, showed that there were some twenty-one loci (plus the single *mgl* locus which also determines type S motility) at which mutations occur that reduce or abolish A-type, single cell, gliding motility. The great majority (approximately 90%) of all A^- isolates were non-motile under all tested conditions, and were assigned to sixteen *agl* (absence of gliding) loci. The remainder of the mutants, designated *cgl* (contact or conditional gliders), would glide when they had been in contact with wild-type cells, *agl* cells, or *cgl* mutants at another locus. Five separate *cgl* loci, B – F, were identified. *Cgl* mutants at loci D and E were motile at lowered temperature (25°C) or on low concentrations of nutrients. Mutual stimulation of gliding occurred between *cgl* mutants at different loci, and *cgl* C, D or E, but not *cgl* B or F, could be stimulated to glide by contact with wild-type or *agl* cells which had been inactivated by formaldehyde.

It was concluded that the sixteen *agl* loci control movement responses i.e. mutants may be defective in the actual mechanism of gliding, or unable to detect or respond to movement stimuli. In contrast, mutants at the five *Cgl* loci fail to produce, but are fully capable of responding to, one of five different individual movement stimuli. The stimuli were shown to be either contact-transmitted, or to operate only over very short intercellular distances. *Cgl* D mutants were subsequently shown by Suva (1979) to be capable of gliding on special surfaces, such as dialysis membrane or polystyrene, even though not moving on agar. Furthermore Helman (1979) reported that outer membrane material of vesicular form would stimulate motility in two classes of *cgl* mutants. Polar filaments and blebs of such material had earlier been observed on living cells of *M. xanthus* by Suva (1978) who reported that the polar filaments showed slow elongation but rapid contraction. Osmotic shock caused reversible disappearance of the filaments. However it is not clear if such retractable polar filaments (? fimbriae) are associated with A- or S-type movement. It is obviously highly desirable for an understanding of type A, single-cell, motility that the gene products of the *agl* and *cgl* loci be identified and their individual functions be determined, by biochemical and immunological techniques.

Gliding of groups of cells, type S motility, is not shown by mutants at the single *mgl* locus, which also determines type A motility. S motility is absent, but type A is present, in mutants at nine loci named *sgl* A – G (social gliding) and *tgl* (transient gliding). *Sgl* mutants do not glide in cell groups but *tgl* mutants can be transiently stimulated to do so by contact with wild-type or *sgl* cells. The loci of system S, both *sgl* and *tgl,* control cell piliation. S^- cells are non-piliated, and the pili have been shown to be needed for the group interaction between cells which leads to social gliding, rather than the actual motility. *Tgl* mutants become transiently piliated while transiently group-motile (Kaiser, 1979).

EDTA has been reported by Morandi (1979) to remove the pili from S^+ cells and to reduce motility. It is not clear whether A type motility is also affected by this treatment. In contrast to the above description of the relationship between pili and S type motility, Morandi (1979) has also reported that there are *three* phenotypic classes of S^- cells. These are those giving (1) heaped colonies, non-dispersed growth in liquid, and heavily piliated cells; (2) flat colonies, dispersed growth in liquid, and non-piliated cells; and (3) fringed colonies, some residual motility, dispersed growth in liquid, and piliated cells. Thus it is not clear whether the *sgl* and *tgl* mutants which have been analysed (Hodgkin and Kaiser, 1977, 1979*a,b;* Kaiser, 1979) were all of Morandi's flat colony type. If so the heaped and fringed S^- motility types might well be defective in the actual mechanism of group motility, rather than merely the inter-cell group interaction via pili. Approximately two thirds of the non-piliated S^- mutants are unable to form normal fruiting bodies, suggesting that motility system S is involved in aggregation during fruiting (A^- isolates will all fruit normally). However since one third of S^- non-piliated mutants do fruit, there is by no means a direct relationship between group motility and fruiting ability.

There remain several unanswered questions connected with type A and type S motility. For example, is system A in any way connected with the longitudinal bundles of cytoplasmic filaments (rhapidosomes) seen below cell membranes

(Avery, 1979) which may be analogous to similar structures apparently defective in some non-motile strains (Burchard, Burchard & Kloetzel, 1977)? It is unlikely that these filaments are pili precursors since many of the fringed colony type of S^- cells have reduced or no rhapidosomes and yet are piliated (Avery, 1979).

What are we to make of swarmer versus non-swarmer phase variation in *M. xanthus,* and the fact that in the non-swarmer phase motility (but whether of A and/or S type we do not as yet know) is curtailed?

Do the extensible blebs and polar fimbriae (pili), the latter apparently associated only with system S motility, relate in any way to the rotary assemblies in the cell envelope which, from work with *Cytophaga* and *Flexibacter* Pate & Chang (1979) have suggested may be responsible for gliding motility?

Are the actual mechanisms of system A and system S gliding motility very different, with only one, or both, operating under different sets of environmental conditions? Is it system A and/or system S motility which is responsible for the apparent chemotaxis towards adenosine monophosphate (AMP) or cyclic guanosine monophosphate (cGMP), but not cyclic adenosine monophosphate (cAMP), as described by Ho & McCurdy (1979) and by Shimkets, Dworkin & Keller (1979)?

How are type A and type S motility involved in the rhythmic waves seen in time-lapse films to pass through swarms of Myxobacteria (Kuhlwein & Reichenbach, 1965)? Procrustean attempts have been made to interpret these and related phenomena, such as aggregation in fruiting, in terms of the cAMP pulses and phosphodiesterase activity involved in cellular slime mould morphogenesis (Campos & Zusman, 1975; Parish, Wedgwood & Herries, 1976; Jones, 1978). However, the *apparent* chemotaxis of *M. xanthus* towards cGMP and AMP but *not* cAMP is now believed to be an artifact so that the analogy may be dangerously misleading (Ho & McCurdy, 1979; Cleary & Foster, 1978).

Do the drugs which have been shown to abolish or reduce gliding motility in *Cystobacter* (Heumann & Kühlwein, 1979) affect the A or S motility systems of *M. xanthus?*

How are the various motility genes of *M. xanthus* arranged on the genetic map? So far there is only fragmentary transductional evidence available indicating, for example, that the *cgl* B, C and D genes of system A, and the *tgl* locus of system S, are all linked to the RNA polymerase (Rif^r) locus (Sodergren, 1979).

The answers to these and other questions concerning the motility systems of *M. xanthus* should not be long forthcoming, judging by the recent rate of progress in the field. Fig. 3 gives an outline of the genetic control of motility in *M. xanthus*.

Fruiting body formation

In prophetic papers published in 1962 and 1963 respectively McVittie, Messik & Zahler and Leadbetter isolated non-fruiting mutants of *M. xanthus*. This work was not, alas, taken up by others until some one and a half decades later (Hagen, Bretscher & Kaiser, 1978), a clear example of a discovery ahead of its time. McVittie *et al.* (1962) identified two (and a possible third) classes of non-fruiting mutants. Of their two main classes one retained, while the other had lost, the ability to form myxospores under fruiting conditions i.e. the former class was aggregation-defective. Mixed cultivation of representatives of the two main classes led to cooperative fruiting on appropriate medium. In this same paper and a related note (McVittie & Zahler, 1962) indirect evidence was also presented for the possible involvement of chemotaxis in aggregation and fruiting. Fruiting bodies were formed above a membrane in a spatial arrangement greatly influenced by underlying fruiting bodies, and the effect could be shown to be other than merely mechanical. Support for this conclusion also comes from another early report using a flow technique in submerged culture (Fleugel, 1963, 1964). Elasticotaxis, the directed movement along stress lines in the substratum, including perhaps dried slime trails, may also play a natural role in fruiting (Stanier, 1942; Dworkin, personal communication), although this may be a guidance phenomenon rather than a true taxis (see Dunn, the first chapter in this book).

More recent work by Shimkets *et al.* (1979) and by Ho & McCurdy (1979) using special apparatus to maintain long-lived concentration gradients, has shown that *M. xanthus* is, apparently, chemotactic towards AMP and cGMP, but not cAMP. However these effects have more recently been shown to be, most probably, an artifact of elasticotaxis (M. Dworkin, personal communication). It is not clear whether this particular effect involves motility systems A and/or S, and whether it operates only in outward swarming during vegetative growth or during inward aggregation preceding fruiting. Certainly the spatial positioning

Fig. 3. Systems of motility in *M. xanthus*. (After Hodgkin & Kaiser (1977), modified.)

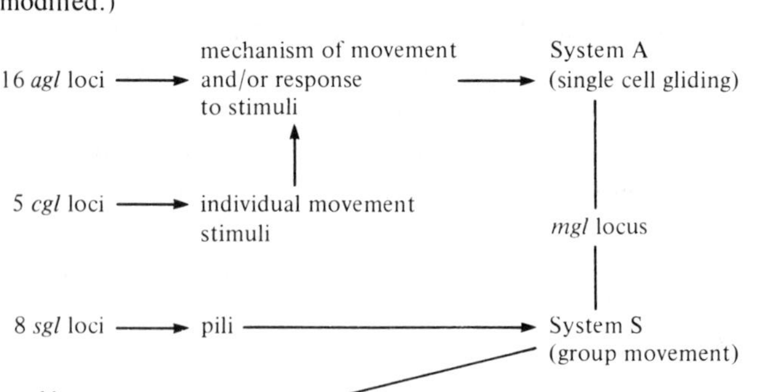

of fruiting bodies on the surface of solid medium is consistent with aggregation to foci under the influence of chemotaxis (Fluegel, 1965; Parish *et al.*, 1976). *M. xanthus* phosphodiesterase, active on cGMP but not cAMP, has also been implicated in aggregation and fruiting (McCurdy, Ho & Dobson, 1978).

In their analysis of some forty non-fruiting mutants, Hagen *et al.* (1978) distinguished three general phenotypic classes. These were, firstly, those arrested at some intermediate stage of normal fruiting. Secondly there were mutants with abnormal motility, such as S^- motility mutants, and thirdly there were mutants producing abnormal structures not seen in normal fruiting ('deranged fruiters'). Fourteen mutants blocked in normal fruiting were tested in pairwise combinations for their ability to form intact fruiting bodies, or solely free-lying myxospores. About half the fruiting-defective mutants were like isolate DK510 in that they could not be induced to form even free myxospores when mixed with wild-type cells. However about 50% of this subclass would still respond to glycerol induction of myxospore formation.

The other half of the fruiting-defective mutants were like isolate DK563 in that they formed myxospores when mixed with wild-type cells, and also gave glycerol-induced myxospores. Tested in pairwise combinations sixteen such mutants fell into three subgroups which would cooperatively form myxospores, though not, in most cases, fruiting bodies i.e. aggregation was still defective. One of these three subgroups had a stable tan colonial phenotype.

Among their isolates Hagen, Bretscher & Kaiser (1978) noted some, like isolate DK642, which actually *inhibited* the fruiting and myxosporulation of the wild type. Such fruiting-inhibitory mutants had earlier been reported by Leadbetter (1963). Conversely, Hemphill & Zahler (1968) isolated a superfruiting uracil auxotroph which formed fruiting bodies, and even induced fruiting in admixed wild-type cells on complete medium where, normally, only vegetative growth would occur. Another superfruiter was also isolated by Rosenberg, Filer, Zafriti & Kindler (1973), and showed an alteration in the feedback inhibition control of aspartokinase activity.

Finally, from an elegant analysis of temperature-sensitive non-fruiting mutants Morrison & Zusman (1979) were able to show that in normal fruiting, aggregation and mound formation on one hand, and myxosporulation on the other, are largely independent pathways. Both pathways have essential early steps, blockage of which has a delayed phenotypic effect, but the two pathways can be blocked separately (Foster, 1980).

The following is an outline scheme for fruiting-body formation in *M. xanthus*, based on my interpretation of the published papers. For normal fruiting to occur there are three factors which must all be satisfied. Firstly there must be a solid surface. Secondly there must be a sufficient cell density, and thirdly there must at the same time be a nutritional deprivation (M. Dworkin, personal communication). On starvation, not only for the essential, required, amino acids but also for carbon, nitrogen or phosphorus source (Manoil, 1978; Manoil & Kaiser, 1980*b*) the fruiting response is initiated. Guanosine tetraphosphate and guanosine pentaphosphate seem to be involved at this starvation-initiation stage (Manoil & Kaiser, 1980*a,b*). Effects operating at the level of a decrease in aspartokinase activity also seem to be important in causing amino-acid starvation, as evidenced by the altered regulation of this activity in a superfruiting mutant (Rosenberg *et al.,* 1973) and the fact that some amino acids will inhibit fruiting under otherwise suitable starvation conditions (Leadbetter, 1963; Dworkin, 1963; Hemphill & Zahler, 1968; Rosenberg *et al.,* 1973; Campos & Zusman, 1975; Manoil & Kaiser, 1980*b*).

The finding that under marginal conditions adenosine diphosphate (ADP), cAMP, and to a lesser extent other adenine compounds, will stimulate fruiting set off a largely unrewarding search for a central role of cAMP in fruiting. Since ADP is a strong inhibitor of aspartokinase (Rosenberg *et al.,* 1973), and a histidine auxotroph would fruit on a low-histidine medium if purines were also present (Hemphill & Zahler, 1968), it seems likely that the suggestion of Kaiser, Manoil & Dworkin (1979) that ADP, cAMP and related compounds exert their stimulatory effects on fruiting via 5-phospho-d-ribosyl-pyrophosphate (PRPP) feedback inhibition, leading indirectly to amino-acid starvation, is correct (Cleary & Foster, 1978). Glycine has been shown to reverse the stimulation of fruiting caused by adenine compounds (Manoil & Kaiser, 1980*c*).

The finding by Rudd & Zusman (1979) that about 20% of Rif^r isolates show heterogeneous derangements of normal fruiting suggests that RNA polymerase or promoter modifications are involved at various steps controlling the transcription of genes for aggregation and mound formation, on one hand, and myxosporulation on the other.

The process of aggregation and mound formation proceeds along one developmental pathway. Aggregation obviously involves motility system S, since most, but not all, S$^-$ mutants (*tgl*$^-$ and *sgl*$^-$) are non-fruiters. The polar fimbriae may in some way play a role in cell-to-cell aggregation or signal transmission. The myxobacterial haemagglutinin (protein H) may, perhaps, be involved in adhesion of aggregating cells (Cumsky & Zusman, 1979) although it has, as yet, not been proved to have such an intercellular function. Although, as mentioned earlier, *M. xanthus* has been reported to be positively chemotactic towards cGMP or AMP, and phosphodiesterase antagonizes this effect, it is by no means clear whether it is the S system of motility, operative in fruiting, which is affected, or the A system, which seems not to be involved in fruiting (Ho & McCurdy, 1979; Shimkets *et al.,* 1979).

In parallel with the sequence of events constituting the aggregation pathway of the fruiting process there occurs a series of other events leading, eventually, to the production of mature fruiting-body myxospores. Preferential lysis of those cells in the wild-type population which are in the tan, non-swarmer, phase seems to be a precondition for successful natural myxosporulation (Wireman & Dworkin, 1977; Wireman, 1978; Janssen & Dworkin, 1979). There is also, probably, a prerequisite for a critical mass of aggregated cells for the initiation of natural myxosporulation (Hemphill & Zahler, 1968; Parish *et al.*, 1976; Wireman & Dworkin, 1977). Adenosine has recently been shown to be the density-dependent signal for natural myxosporulation (Shimkets & Dworkin, 1980). Synthesis followed by assembly of protein S onto the outer layer of the fruiting body myxospores is one well characterized biochemical marker in this pathway (Inouye, Inouye & Zusman, 1979*a,b*). Numerous soluble and membrane proteins undergo changes in synthesis during fruiting (Orndorff & Dworkin, 1980; Inouye *et al.*, 1979*b*), and it has been shown that many of the later changes do not occur in the presence of excess L-methionine, a specific inhibitor of fruiting. The naturally formed myxospores inside fruiting bodies contain protein S, have a triple-layered wall, and no excess internal membranes. In contrast, myxospores formed on glycerol induction of non-starving vegetative cells have no protein S, and will not assemble it onto their surface if it is supplied, have a one-layered wall, and convoluted internal membranes (Inouye *et al.*, 1979*a,b*). Evidently in glycerol induction of cellular morphogenesis something very different, and probably rather simpler, occurs compared with the more leisurely programming of natural myxospore formation within fruiting bodies. A biochemical and electron-microscopical examination of the formation of myxospores in response to other stimuli, such as phenethyl alcohol (Burchard & Parish, 1975) or methionine starvation (Witkin & Rosenberg, 1970) is obviously highly desirable to determine whether these treatments produce myxospores which are more natural than those produced by glycerol and related inducers.

The above ideas are summarized in Fig. 4, which also indicates the expected phenotypes of mutants blocked at various steps in the hypothetical fruiting scheme.

There remain several unresolved questions, concerning fruiting, the answers to which will, undoubtedly, be obtained soon, mostly in laboratories in the USA. Firstly, fruiting-defective mutants might well be investigated for their possible phenotypic suppression by analogues (Manoil & Kaiser, 1980*a,b,c*). Leadbetter (1963) showed that some non-fruiting, non-myxosporulating, isolates would fruit in the presence of β-2-thienyl alanine. Thus it will be useful to test non-fruiting mutants for their ability to fruit in the presence of known stimulators of normal, wild-type, fruiting, such as *p*-fluorophenylalanine (Hemphill & Zahler, 1968), threonine (Rosenberg *et al.*, 1973; Campos & Zusman, 1975), adenine compounds (Campos & Zusman, 1975), or cGMP-specific phosphodiesterase (McCurdy *et al.*, 1978).

It is clear that the fruiting-defective mutants are being investigated for alterations in the patterns of synthesis of soluble and membrane proteins, including haemagglutinin, which occur during normal fruiting. Work has recently been carried out by Inouye *et al.* (1979*a,b*), Morrison & Zusman (1979), and Cumsky & Zusman (1979), using gel and immunological techniques. Cloning of specific myxobacterial developmental genes governing the production of products such as proteins S and H (haemagglutinin) is under way (Faulds & Zusman, 1979).

In the field of genetics much use will obviously be made of transposon Tn5 mutagenesis, derived from infecting phage P1Km (Kuner, 1979). Detailed mapping of the morphogenetic genes of *M. xanthus* is desirable, perhaps using long-range R factor-mediated chromosome mobilization (Parish, 1975), as has been applied so successfully to *Pseudomonas, Rhizobium* and *Acinetobacter* species. The promiscuous plasmid R68.45 may be useful in this context (McCann, 1980). The identification of specific gene products, whether structural or regulatory, for all known fruiting, motility systems A and S, polar fimbriae, autolysis, aggregation, and myxosporulation genes is one end goal. As with *Bacillus subtilis spo* genes this may not be an easy task.

Fig. 4. Outline scheme for fruiting body formation and myxosporulation in *M. xanthus*. Block at stage: (1), $Agg^-Myx^-Ind^-$; (2), $Agg^-Myx^-Ind^+$; (3), $Agg^-Myx^+Ind^+$; (4), $Agg^+Myx^-Ind^+$; (5), $Agg^+Myx^+Ind^-$. Unable to form: aggregation mounds, Agg^-; normal (fruiting body) type myxospores, Myx^-; myxospores on glycerol induction, Ind^-. GtP, guanosine tetraphosphate; GpP, guanosine pentaphosphate.

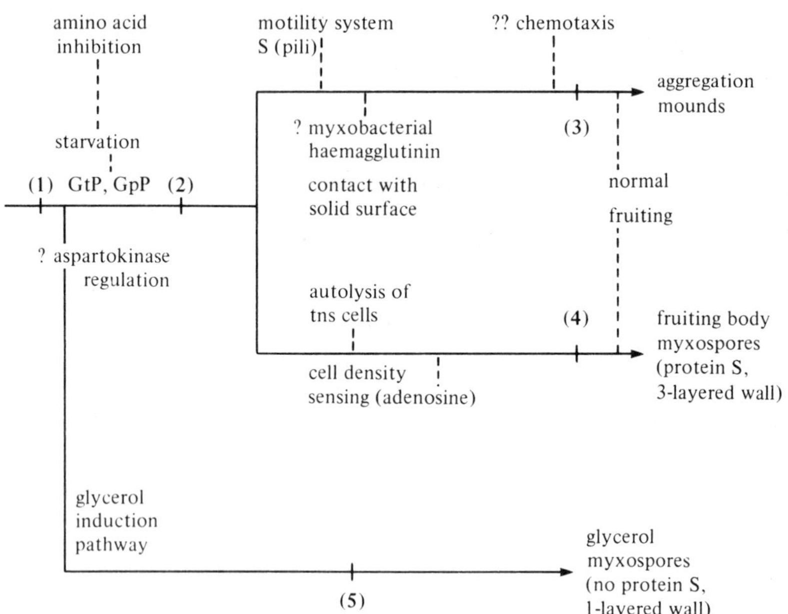

Conclusion

I hope that in this review I have brought to the attention of biologists who are not bacteriologists the impressive recent progress which has been made with Myxobacteria in the understanding of the genetics, but not yet the actual mechanism, of gliding motility and of the complex processes involved in fruiting body formation. Movement towards food microbes, adherence together in moving swarms, and aggregation to foci in fruiting body formation are all phenomena which in Myxobacteria may depend upon various types of chemotaxis and intercellular signalling but are not yet proven to do so. It may well be that these processes, in a prokaryote, are rather different in detail, though superficially similar, to those shown by, especially, the cellular, and to a somewhat lesser extent the true plasmodial, types of slime moulds. Most of the myxobacterial work has been carried out with *M. xanthus* which forms the simplest, tumulus-like, fruiting body. Other Myxobacteria such as *Stigmatella aurantiaca* and *Chondromyces apiculatus,* both of which form tree-like, elaborately sculptured, fruiting bodies with large cysts enclosing dormant swarms of cells, deserve, and will certainly receive, increased attention (Qualls, Stephens & White, 1978).

I wish to thank, in the following approximate geographical order, all who have helped my understanding and enjoyment of the Myxobacteria by their tuition, advice, discussions, encouragement and friendship – Hans Reichenbach, Klaus Gerth, Klaus Grimm, Hans Kuhlwein, Klaus Herdrich, Gerhard Ruckert, Marty Dworkin, John Wireman, Palmer Rogers, Doris Overby, Beverly Smith, Paul Orndorff, David Weisberg, Mary Turnblom, June Blalock, Gary Jannsen, Larry Shimkets, Howard Parish, Robert Burchard, Martin Jones, Irene Napier, Howard Foster, Kate McCann, Wendy Cleary, Diane Hawkins, Peter Miller, Richard Brown, Diana Fox, Sue Katz, Del Shankel.

I am extremely grateful to M.A.S.U.A. for funding my visit to K.U. and other Mid-Western Universities, and to Professor D. M. Shankel for his hospitality at K.U. which made possible the invaluable period of academic peace needed to write this article. Professor Martin Dworkin very kindly read, and provided much-appreciated constructive criticism of, a first draft of this review.

Research on *M. xanthus* in my own laboratory has been supported by S.R.C. grants GR/A 76818 and GR/A 68837, and by funds from the University of East Anglia.

References

Avery, L. (1979). The relationship between rhapidosomes and gliding motility in *M. xanthus. Abstracts of the 6th International Conference on Myxobacteria,* Asilomar, Pacific Grove, California.

Bretscher, A. P. & Kaiser, D. (1978). Nutrition of *Myxococcus xanthus,* a fruiting myxobacterium. *Journal of Bacteriology,* **133,** 763–8.

Brockman, E. R. & Todd, R. L. (1974). Fruiting myxobacters as viewed with a scanning electron microscope. *International Journal of Systematic Bacteriology,* **24,** 118–24.

Burchard, A. C., Burchard, R. P. & Kloetzel, J. A. (1977). Intracellular, periodic structures in the gliding bacterium *Myxococcus xanthus. Journal of Bacteriology,* **132,** 666–72.

Burchard, R. P. (1970). Gliding motility mutants of *Myxococcus xanthus. Journal of Bacteriology,* **104,** 940–7.

Burchard, R. P. (1974). Studies on gliding motility in *Myxococcus xanthus. Archiv für Mikrobiologie* **99,** 271–80.

Burchard, R. P., Burchard, A. C. & Parish, J. H. (1977). Pigmentation phenotype instability in *Myxococcus xanthus. Canadian Journal of Microbiology,* **23,** 1657–1662.

Burchard, R. P. & Dworkin, M. (1966). Light-induced lysis and carotogenesis in *Myxococcus xanthus. Journal of Bacteriology,* **91,** 535–45.

Burchard, R. P. & Parish, J. H. (1975). Mutants of *Myxococcus xanthus* insensitive to glycerol-induced myxospore formation. *Archiv für Mikrobiologie,* **104,** 289–92.

Campos, J. M. & Zusman, D. R. (1975). Regulation of development in *Myxococcus xanthus:* effect of 3′:5′-cyclic AMP, ADP, and nutrition. *Proceedings of the National Academy of Sciences of the U.S.A.,* **72,** 518–22.

Cleary, W. G. & Foster, H. A. (1978). Inhibition of vegetative growth and fruiting body formation in *Myxococcus xanthus. FEMS Microbiology Letters,* **4,** 265–69.

Cumsky, M. & Zusman, D. R. (1979). Myxobacterial hemagglutinin: a development-specific lectin of *Myxococcus xanthus. Proceedings of the National Academy of Sciences of the U.S.A.,* **76,** 5505–9.

Dobson, W. J., McCurdy, H. D. & MacRae, T. H. (1979). The function of fimbriae in *Myxococcus xanthus.* II. The role of fimbriae in cell-cell interaction. *Canadian Journal of Microbiology,* **25,** 1359–72.

Dworkin, M. (1963). Nutritional regulation of morphogenesis in *Myxococcus xanthus. Journal of Bacteriology,* **86,** 67–72.

Dworkin, M. (1966). Biology of the myxobacteria. *Annual Reviews of Microbiology,* **20,** 75–106.

Dworkin, M. (1972). The myxobacteria: new directions in studies of procaryotic development. *CRC Critical Reviews in Microbiology,* **1,** 435–52.

Dworkin, M. (1973*a*). Cell-cell interactions in the myxobacteria. In *Symposium No. 23 of the Society of General Microbiology 'Microbial Differentiation',* pp. 125–142. London: Cambridge University Press.

Dworkin, M. (1973*b*). The Myxobacterales {(fruiting) Myxobacterales}. *Handbook of Microbiology,* vol. **1,** *Organismic Microbiology,* ed. A. I. Laskin & H. A. Lechevalier, pp. 191–202. Cleveland: CRC Press.

Faulds, D. & Zusman, D. (1979). Cloning development specific genes from *M. xanthus. Abstracts of the 6th International Conference on Myxobacteria,* Asilomar, Pacific Grove, California.

Fluegel, W. (1963). Fruiting chemotaxis in *Myxococcus fulvus* (Myxobacteria). *Proceedings of Minnesota Academy of Sciences,* **30,** 120–2.

Fluegel, W. (1964). Induced fruiting in Myxobacteria. *Proceedings of Minnesota Academy of Sciences,* **31,** 114–15.

Fluegel, W. (1965). Fruiting body populations of *Myxococcus fulvus* (Myxobacterales). *Growth,* **29,** 183–91.

Foster, H. A. (1980). Developmental mutants of the fruiting myxobacterium *Myxococcus xanthus. Heredity,* **43** (in press).

Hagen, D. C., Bretscher, A. P. & Kaiser, D. (1978). Synergism between morphogenetic mutants of *Myxococcus xanthus. Developmental Biology,* **64,** 284–96.

Helman, T. (1979). Correction of a motility defect by membrane vesicles of

Myxococcus xanthus. Abstracts of the 6th International Conference on Myxobacteria, Asilomar, Pacific Grove, California.

Hemphill, H. E. & Zahler, S. A. (1968). Nutritional induction and suppression of fruiting in *Myxococcus xanthus. Journal of Bacteriology,* **95,** 1018–23.

Heumann, H. G. & Kühlwein, H. (1979). Effects of drugs that influence eucaryotic motile processes on motility of *Cystobacter fuscus* (Myxobacterales). *Zeitschrift für allgemeine Mikrobiologie,* **19,** 547–52.

Ho, J. & McCurdy, H. D. (1979). Demonstration of positive chemotaxis to cyclic GMP and 5′-AMP in *Myxococcus xanthus* by means of a simple apparatus for generating practically stable concentration gradients. *Canadian Journal of Microbiology,* **25,** 1214–18.

Hodgkin, J. & Kaiser, D. (1977). Cell-to-cell stimulation of movement in nonmotile mutants of *Myxococcus. Proceedings of the National Academy of Sciences of the U.S.A.,* **74,** 2938–42.

Hodgkin, J. & Kaiser, D. (1979*a*). Genetics of gliding motility in *Myxococcus xanthus* (Myxobacterales): genes controlling movement of single cells. *Molecular and General Genetics,* **171,** 167–76.

Hodgkin, J. & Kaiser, D. (1979*b*). Genetics of gliding motility in *Myxococcus xanthus* (Myxobacterales): two gene systems control movement. *Molecular and General Genetics,* **171,** 177–91.

Inouye, M., Inouye, S. & Zusman, D. R. (1979*a*). Biosynthesis and self-assembly of protein S, a development-specific protein of *Myxococcus* xanthus. *Proceedings of the National Academy of Sciences of the U.S.A.,* **76,** 209–13.

Inouye, M., Inouye, S. & Zusman, D. R. (1979*b*). Gene expression during development of *Myxococcus xanthus:* pattern of protein synthesis. *Developmental Biology,* **68,** 579–91.

Janssen, G. R. & Dworkin, M. (1979). Control of developmental autolysis in *M. xanthus* by self-generated signals. *Abstracts of the 6th International Conference on Myxobacteria,* Asilomar, Pacific Grove, California.

Jones, M. V. (1978). Inhibition of fruiting body formation in *Myxococcus xanthus* by cytokinins. *FEMS Microbiology Letters,* **3,** 147–9.

Kaiser, D. (1978). Genetics of cell interactions in Myxobacteria. *Birth Defects: Original Article Series,* **14,** 391–9.

Kaiser, D. (1979). Social gliding is correlated with presence of pili in *Myxococcus xanthus. Proceedings of the National Academy of Sciences of the U.S.A.,* **76,** 5952–6.

Kaiser, D., Manoil, C. & Dworkin, M. (1979). Myxobacteria: cell interactions, genetics, and development. *Annual Reviews of Microbiology,* **33,** 595–639.

Kühlwein, H. & Reichenbach, H. (1965). Schwarmentwicklung und Morphogenese bei Myxobakterien *Archangium – Myxococcus – Chondrococcus – Chondromyces.* Film C893/1965. Göttingen, West Germany: Institut für den Wissenschaftlichen Film.

Kuner, J. M. (1979). Use of transposon Tn5 in *Myxococcus. Abstracts of the 6th International Conference on Myxobacteria,* Asilomar, Pacific Grove, California.

Leadbetter, E. R. (1963). Control of growth and morphogenesis in some *Myxococcus* species. *Nature, London,* **200,** 1127–8.

McCann, A. K. (1980). Plasmid-mediated UV protection in *Myxococcus xanthus. Photobiology Bulletin,* **1,** 229–30.

McCurdy, H. D., Ho, J. & Dobson, W. J. (1978). Cyclic nucleotides, cyclic nucleotide phosphodiesterase, and development in *Myxococcus xanthus*. *Canadian Journal of Microbiology*, **24**, 1475–81.

McVittie, A., Messik, F. & Zahler, S. A. (1962). Developmental biology of *Myxococcus*. *Journal of Bacteriology*, **84**, 546–51.

McVittie, A. & Zahler, S. A. (1962). Chemotaxis in *Myxococcus*. *Nature, London*, **194**, 1299–300.

Manoil, C. & Kaiser, D. (1980*a*). Accumulation of guanosine tetraphosphate and guanosine pentaphosphate in *Myxococcus xanthus* during starvation and myxospore formation. *Journal of Bacteriology*, **141**, 297–304.

Manoil, C. & Kaiser, D. (1980*b*). Guanosine pentaphosphate and guanosine tetraphosphate accumulation and induction of *Myxococcus xanthus* fruiting body development. *Journal of Bacteriology*, **141**, 305–15.

Manoil, C. & Kaiser, D. (1980*c*). Purine-containing compounds, including cyclic adenosine 3′,5′-monophosphate, induce fruiting of *Myxococcus xanthus* by nutritional imbalance. *Journal of Bacteriology*, **141**, 374–7.

Manoil, C. (1978). The induction of fruiting body formation in *Myxococcus xanthus*. *Abstracts of the 5th International Conference on Myxobacteria*, Wayzata, Minnesota.

Morandi, D. (1979). Genetic control of S-motility in *M. xanthus*. *Abstracts of the 6th International Conference on Myxobacteria*, Asilomar, Pacific Grove, California.

Morrison, C. E. & Zusman, D. R. (1979). *Myxococcus xanthus* mutants with temperature-sensitive, stage-specific defects: evidence for independent pathways in development. *Journal of Bacteriology*, **140**, 1036–42.

Orndorff, P. E. & Dworkin, M. (1980). Separation and properties of the cytoplasmic and outer membranes of vegetative cells of *Myxococcus xanthus*. *Journal of Bacteriology*, **141**, 914–27.

Parish, J. H. (1975). Transfer of drug resistance to Myxococcus from bacteria carrying drug-resistance factors. *Journal of General Microbiology*, **87**, 198–210.

Parish, J. H. (1979). *Myxobacteria in Developmental Biology of Prokaryotes* (ed. J. H. Parish), pp. 227–53. Oxford: Blackwell.

Parish, J. H., Wedgwood, K. R. & Herries, D. G. (1976). Morphogenesis in *Myxococcus xanthus* and *Myxococcus virescens* (Myxobacterales). *Archives for Microbiology*, **107**, 343–51.

Pate, J. L. & Chang, L.-Y. E. (1979). Evidence that gliding motility in prokaryotic cells is driven by rotary assemblies in the cell envelopes. *Current Microbiology*, **2**, 59–64.

Qualls, G. T., Stephens, K. & White, D. (1978). Light-stimulated morphogenesis in the fruiting myxobacterium *Stigmatella aurantiaca*. *Science, New York*, **201**, 444–5.

Rosenberg, E., Filer, D., Zafriti, D. & Kindler, S. H. (1973). Aspartokinase activity and the developmental cycle of *Myxococcus xanthus*. *Journal of Bacteriology*, **115**, 29–34.

Rosenberg, E., Keller, K. H. & Dworkin, M. (1977). Cell density-dependent growth of *Myxococcus xanthus* on casein. *Journal of Bacteriology*, **129**, 770–7.

Rudd, K. & Zusman, D. R. (1979). Rifampin-resistant mutants of *Myxococcus xanthus* defective in development. *Journal of Bacteriology*, **137**, 295–300.

Shimkets, L. & Dworkin, M. (1981). *Journal of Bacteriology* (in press).

Shimkets, L., Dworkin, M. & Keller, K. H. (1979). A method for establishing stable concentration gradients in agar suitable for studying chemotaxis on a solid surface. *Canadian Journal of Microbiology,* **25,** 1460–7.

Sodergren, E. (1979). Genetic analysis of stimulation in *Myxococcus xanthus. Abstracts of the 6th International Conference on Myxobacteria,* Asilomar, Pacific Grove, California.

Stanier, R. Y. (1942). A note on elasticotaxis in Myxobacteria. *Journal of Bacteriology,* **44,** 405–12.

Sudo, S. Z. & Dworkin, M. (1973). Comparative biology of prokaryotic resting cells. *Advances in Microbial Physiology,* **9,** 153–224.

Suva, R. (1978). Surface filaments of *M. xanthus. Abstracts of the 5th International Conference on Myxobacteria,* Wayzata, Minnesota.

Suva, R. (1979). CglD motility mutants in *Myxococcus xanthus. Abstracts of the 6th International Conference on Myxobacteria,* Asilomar, Pacific Grove, California.

Wireman, J. W. (1978). Myxospore differentiation during the development cycle of *Myxococcus xanthus. Abstracts of the 5th International Conference on the Biology of the Myxobacteria,* Wayzata, Minnesota.

Wireman, J. W. & Dworkin, M. (1975). Morphogenesis and developmental interactions in myxobacteria. *Science, New York,* **189,** 516–22.

Wireman, J. W. & Dworkin, M. (1977). Developmentally induced autolysis during fruiting body formation by *Myxococcus xanthus. Journal of Bacteriology,* **129,** 796–802.

Witkin, S. S. & Rosenberg, E. (1970). Induction of morphogenesis by methionine starvation in *Myxococcus xanthus:* polyamine control. *Journal of Bacteriology,* **103,** 641–9.

Zusman, D. R., Krotoski, D. M. & Cumsky, M. (1978). Chromosome replication in *Myxococcus xanthus. Journal of Bacteriology,* **133,** 122–9.

INDEX

acetate, *Paramecium* and, 125
N-acetylglucosamine: *E. coli* and, 145; *Physarum* and, 118, 119, 120
actin, in leucocytes, 32
actomyosin system, in slime-mould amoebae, 104
adaptation: in bacteria, 140–1, 148, 151; in *Dictyostelium,* 105, 106; kinesis showing, 11–13
adenylate cyclase, in *Dictyostelium,* 101–2, 102–3
adhesion: of leucocytes to endothelium of blood vessels, 29; chemokinesis and, 36–7; chemotaxis and, 38–9; locomotion and, 33–5
adhesiveness, of substratum, 10, 21
albumin, *see* serum albumin
Allomyces, 124, 135; chemotaxis in gametes and zygotes of, 117, 127, 128; swimming characteristics of, 128–9, 130–1
amino acids: as attractants for *Allomyces* zygotes, 127, and for *E. coli,* 143, 145; required by Myxobacteria, 156
ammonium ion, *Dunaliella* and, 124
amoeboid movement, sol⪯gel transformation in, 32
cAMP, 91; as aggregation signal in *Dictyostelium,* 101–5, 107; cytotaxins and, 42
antibodies: inhibiting gliding motility in *Myxococcus,* 158–9; receptors for, on leucocytes, 66, 67, 68
antigens, as attractants for antibody-coated leucocytes, 66–7
arachidonic acid derivatives: as chemotactic agents, 53; released in inflammatory reactions, 54
arginine, *Chlamydomonas* and, 124
aspartate, *E. coli* and, 143, 145
aspartokinase, in *Myxococcus,* 163, 164
ATPase, myosin: in *Dictyostelium* amoebae, 104
attractants: antigens as, for antibody-coated lymphocytes, 66–7; bacteria in gradients of, 31; receptors for, in *E. coli,* 141–2; toxin appearing as, 116
avoiding reaction, 20

Bacillus subtilis, calcium, and locomotion of, 151
bacteria, xi, xii–xiii; as attractant for leucocytes, 54; chemotaxis in, 139–52; folic acid produced by, as attractant for slime-mould amoebae, 89; receptors in, 135; *see also individual species of bacteria*
Badhamia utricularis, 118
bias, in random locomotion, 3, 4, 5, 6
Blepharisma (protozoan), 134
blepharismone, sex attractant of *Blepharisma,* 128
boundary responses, 19–22

C5a peptide from complement protein C5, chemotactic agent, 44, 53, 54, 58–9; receptors for, on leucocytes, 58, 68, 74
$C5a_{des.Arg}$, cytotaxin, 39, 40, 41, 42
calcium: in chemotaxis, 42, 57; in *Dictyostelium,* 103, 104; and motility, 135, 151; Mycobacteria and, 156
carbohydrates (sugars, etc.): attractants, not necessarily metabolized, for *Neocallimastix* zoospores, 123–4, and for *Physarum,* 118–20; receptors for, in *E. coli,* as primarily transport proteins, 142–3
caseins (αs- and β-), as chemotactic agents, 56, 59–60; binding sites for, on neutrophils, 61–2, 64, 65
CAT 1.6.1, cytotaxin, 39, 40, 42, 58
cations, response of *Paramecium* to, 125, 127
cell behaviour, methods for studying, 5, 15
cell density, and fruiting in *Myxococcus,* 163
cell distribution, equilibrium in, 7–8
cell membrane: *E. coli* chemoreceptors involved in transport across, 142; movement of calcium across, in motility, 135; potential across, and chemoresponses in *Paramecium,* 125–6
cell 'memory', 11
chemokinesis, 30; of leucocytes, 35–6, 44–5; *see also* kinesis, klinokinesis *and* orthokinesis
chemokinetic agents, 39–43; native serum albumin, 66; peptides, 75

chemoreceptors (binding sites): coupling of, with effectors, xiv, 144–5, 148, 150; loss and recovery of, 79–85; possible movement of, 14–15, 23; possible recycling of, 77, 85; spatially separated, for directional information, 14; *see also under individual organisms and substances*
chemotactic agents, 39–43; amphipathic and denatured proteins, *see under* proteins; formyl peptides, *see* peptides
chemotaxis, 73, 116; assay of, 6, 116–17; in bacteria, 139–51; in leucocytes, 27, 28, 37–8, 44–5; negative, 92–3, 95, 122–3; in slime moulds, *see under* slime moulds; not a special case of haptotaxis, 38, 46; use of term, 29–30, 31; *see also* taxis
chemotropism index, 37
Chlamydomonas reinhardii, 115, 124, 135
chloramphenicol, and gliding motility in *Myxococcus,* 158
Chondromyces apiculatus (Myxobacterium), 156, 167
cilia, calcium in control of beating of, 135
cobalt, *Chlamydomonas* and, 124
collagen gels, guidance in, 16, 17
complement: activation of, and leucocytes, 38, 53, 54, 55; *see also* C5a peptide
contact guidance, *see* guidance
contact inhibition of locomotion, 23; not found in leucocytes, 28
Crypthecodinum cohnii (dinoflagellate), 124–5, 135
Cutleria multifida (brown alga), 128, 131, 132
cycloheximide: effect of, on *Blepharisma,* 134
cycloxygenase, in inflammatory reactions, 54
Cystobacter, gliding motility in, 161
cytochalasin B, 36, 45, 65
cytoplasm: flow of, in locomotion of leucocytes, 31–2
cytotaxins (substances eliciting chemotactic response), 28, 29; and adhesion of leucocytes, 36, 37, 38–9; variety of effects of, 44

dansyl-asparagine, zoospores of *Phytophthora palmivora* and, 120, 121
demethylase, of *E. coli,* 148
denatured protein, *see* protein
Dendrocoelum lacteum, and light intensity, 12
2-deoxyglucose: *Neocallimastix* zoospores and, 123, 124; *Physarum* and, 119, 120
Dictyostelium discoideum (slime mould), 90; chemotaxis in aggregation of, 94, 95–8, 101–5, 106; chemotaxis in multicellular aggregates of, 108–10; chemotaxis in sexual phase of, 106–7; and folic acid, 90, 91; mutants of, in aggregation behaviour, 96, 98, 100; negative chemotaxis in vegetative amoebae of, 92
Dictyostelium lacteum, D. minutum, 101
Dictyostelium mucuroides, 92, 101
Dictyostelium purpureum, D. rosarium, 101
diffusion coefficient, augmented: as measure of cell motility, 7
diffusion constant: relation of, to random walk, 8–9
dimethyl-β-propiothetin, *Crypthecodinum* and, 125
down-regulation (loss of receptors), 80, 81
Dunaliella tertiolecta (alga), 124

ectocarpen, sex attractant of *Ectocarpus,* 128, 132–3
Ectocarpus siliculosus (brown alga), 128, 132–3, 134
elasticotaxis (movement along stress lines), 162
embryo, cell movement in development of, 23, 24
endocrine systems, down-regulation in, 80
endotoxin, *E. coli* lipopolysaccharide as, 54
environment, types of cell response to: guidance to tensor quantity, simple kinesis to scalar quantity, taxis to vector quantity, 16, 22–4
eosinophils, in hypersensitivity reactions, 43
epidermal growth factor, 80
Escherichia coli: chemoreceptors, 25 classes for attractants, 141–2, and 9 classes for repellants, 145; chemotactic agents released by, 54; chemotaxis in, 139; lipopolysaccharides of, activate complement, 54; mutants of, in flagellar proteins, 144, and transducer proteins, 144, 151, and non-chemotactic, defining 6 chemotactic genes, 144
ethanol, as attractant, 121–2
ethylene diamine tetra-acetic acid, and *Myxococcus,* 158, 160
eukaryotic microbes (motile cells of algae, fungi, myxomycetes, protozoa): chemotaxis in, 115–16, (assays of) 116–17; control of motility in, 135; *see also individual organisms*

fibrillar orientation, guidance as response to, 16, 17
fibrinogen, chemotactic agent, 40
fibroblasts: guidance of, by substratum curvature, 22
fimbriae (pili) of *Myxococcus,* and gliding motility, 159, 160, 161, 164
flagella, bacterial, 140; change in direction of rotation of, at change from swimming to tumbling, 140; genes for, 143, 144; transducers connecting chemoreceptors with motors of, 144–5, 148, 150
p-fluorophenylalanine, stimulates fruiting in *Myxococcus,* 165
folic acid, 91; slime moulds and, 89, 90, 92
formyl-peptides, *see* peptides
fructose: *E. coli* and, 145; *Neocallimastix* zoospores and, 123, 124; *Physarum* and, 118, 119

fucose: *Crypthecodinum* and, 125; *E. coli* and, 142; *Neocallimastix* zoospores and, 123, 124
fucoserraten, sex attractant of *Fucus*, 124, 131–2
Fucus spp. (brown algae), 128, 131–2, 132–3

galactose: *E. coli* and, 142, 145; *Neocallimastix* zoospores and, 123, 124; *Physarum* and, 118, 119, 120
genes: for aggregation process in slime moulds, 98, 100; for gliding motility in *Myxococcus*, 159, 160–1, 166
gliding motility, of *Myxococcus* 'swarmer' cells, 158–61, 162
glucose: *E. coli* and, 142; *Neocallimastix* zoospores and, 123, 124; *Physarum* and, 118, 119
glycerol induction of myxospores, 157, 163, 165
glycoprotein, leucocyte receptor for, 74
cGMP: *Myxococcus* and, 101, 162; in slime moulds, 92, 103, 104
gravitational field, taxis response in, 10
guanosine monophosphate, *see* GMP
guanosine tetraphosphate and pentaphosphate, in fruiting of *Myxococcus*, 164
guanylate cyclase, in slime moulds, 92
guidance (contact guidance), xii, 15–17, 21; assay of, 17–19; elasticotaxis as? 162; as response to a tensor quantity of the environment, 16, 22–4

haemagglutinin, of Myxobacteria, 164, 166
haemoglobin, chemotactic agent after removal of haem and denaturation, 60
haptotaxis, 10, 21; chemotaxis not a special case of, 38, 46
hormone systems, mammalian: methylation of proteins in stimulus-response phenomena in, 150
hydrocarbons, weak sex attractants for brown algae, 131–2, 133
hydrogen ions, repellant for *Phytophthora* zoospores, 123
hydrolytic enzymes, released by neutrophils, 55, 62, 64, 65
hydroxy-eicosatetraenoic acids, leucocytes and, 54–5
5-hydroxytryptophan, *Blepharisma* and, 134

immunological analysis, of gliding motility in *Myxococcus*, 158–9
inflammation: chemotactic agents generated in, 54; leucocytes and, 27, 28, 44, 54
iodoacetate, negative chemotactic agent, 36
ionic concentration, affects cell locomotion, 68
isovaleraldehyde, zoospores of *Phytophthora palmivora* and, 122

kinesis, xii; adapting, 11–13; coexistence of taxis and, 10; as response to a scalar quantity of the environment, 16, 22–4; simple, 6–10; *see also* chemokinesis
klinokinesis, xii, 7, 8, 116; adapting, 12–13, 23; diffusion constant of, 9; distinguished from orthokinesis, 10; not found in leucocytes, 36; in *Paramecium*, 125, 127; use of term, 30, 31

L cells, haptotaxis of, 10, 21
lamellipodia of leucocytes, 15, 31, 32
lectins, in leucocytes, 31
leucocytes, xi, 29–31, 53–5; chemokinesis in, 35–6, 44–5, (and adhesion) 36–7; chemotaxis in, 14–15, 37–8, 44–5, (and adhesion) 38–9; co-existence of kinesis and taxis in, 10, 14; locomotion of, 31–2, 43–4; polymorphonuclear, (chemotaxis in) 73–4, (peptide receptors on) 74–7, 79–85, (uptake of peptides by) 77–9; receptors on, 14, 67–8; types of, showing chemotaxis, 55; *see also* eosinophils, lymphocytes, macrophages, neutrophils, phagocytes
leukotriene B, leucocytes and, 54–5
light intensity, taxis response to gradients of, 10, 12
lipoprotein, low density: receptors for, 80–1, 85
lipoxygenase, in inflammatory reactions, 54
locomotion: of bacteria, swimming and tumbling, 139, 151; chemotaxis in cells capable of, 30; directional, depends on both chemokinetic and chemotactic influences, 39; of leucocytes, 27, 28, 31–2, (and adhesion) 33–5, (selective directional) 43–4, (speed of) 32, 37; non-random, 5–6, (simple kinesis) 6–10, (taxis) 10–22; in polymorphonuclear leucocytes, 83, 85, (rate of) 75, 82; random, 2–5
lymphocytes: antibody-coated, attracted by antigens, 66–7; in virus-induced lesions, 43–4
lysosomal enzymes, released by leucocytes, 28, 44, 45

macrocysts: of *Dictyostelium*, 107; of Myxobacteria, 156, 167
macrophages, 43, 44, 56, 67
magnesium: in chemotaxis, 42; Myxobacteria and, 156
magentic field, taxis response to, 10
malic acid, attractant for fern sperm, 29
maltose: *E. coli* and, 141, 142, 143, 146; *Physarum* and, 118, 119, 120
manganese, attractant for *Chlamydomonas*, 124
mannitol, mannose: *Neocallimastix* zoospores and, 123, 124; *Physarum* and, 118, 119, 120
mechanical stress, guidance as response to, 16, 17
melittin (bee venom), conjugated: as chemotactic agent, 59, 60
metabolism, cell movement affected by stimulants and inhibitors of, 68

metazoan cells: at boundaries, 21; detection of ortho- and klino-kinesis in, 9–10; never random walkers, 3
methionine: *E. coli* and, 143; and fruiting in *Myxococcus,* 165
methyl-α-glucose, *Physarum* and, 118, 119
methyl transferase, 147, 149
methylation of proteins, *see under* proteins
motility: augmented diffusion coefficient as measure of, 7; control of, in eukaryotic microbes, 135; curves of, 10; diminishing to zero in one region of the environment, leads to 'trapping' of all cells in that region, 23; gliding, of *Myxococcus,* 158–61, 162; *see also* locomotion
multifiden, sex attractant of *Cutleria,* 128, 132–3
myosin, in leucocytes, 32
Myxobacteria, xiv, 155–6, 167
Myxococcus xanthus, 156–7, 167; aggregation and myxosporulation in, are independent, 163; fruiting in, 157, 162–6; gliding motility of, 158–61, 162; life cycle of, 155–6; 'swarmer' and 'non-swarmer' colonies of, 158
Myxomycetes, xiv; chemotaxis in, 115, 117–20, 135
myxospores, of Myxobacteria, 156, 165, 166

necrotaxis, 54
Neocallimastix frontalis, rumen organism, zoospore of phycomycete fungus, 116, 123, 124
neutrophils, xii, 33, 43; adhesion of, 34, 44; chemotaxis in, antigen-specific, 66, to denatured proteins, 59–62, to peptides, 41, 42, 53, 54, 55–9, and to surfaces coated with protein, 62–3, 64; movements of, 4, 34, 36, 37, 68

oil pollution of coastal waters, and brown algae, 132
optical properties of slime-mould amoebae: binding of cAMP and, 103, 104; of moving and stationary cells, 94, 95–6, 97
orientation: of amoebae in grex, 108; of leucocytes in gradients, 37, 64, 75, 83–5
orthokinesis, 7, 8, 116; causes of, 35–6; diffusion constant of, 9; distinguished from klinokinesis, 10; in leucocytes, 68; in *Paramecium,* 125, 127; use of term, 31
osmotic shock, and gliding motility in *Myxococcus,* 158, 160

paralysis, of pre-spore zone in slime-mould aggregates, 108–9, 110
Paramecium (protozoan), chemoresponses of, 115, 125–6, 135
peptides, formyl (chemotactic agents), and leucocytes, 41, 42, 53, 54, 55–6, 57; dose-response curve for uptake of, 78; multiple effects of, 28, 39, 40, 55, 76; receptors for, 58, 68, 74–7, 80, (antagonists competing for) 56, 74, 78; release of proteases in response to, 65
persistence (internal bias), 3, 4, 5
pH: at binding of cAMP in *Dictyostelium,* 103; and cell movement, 68; *Paramecium* and, 125, 126
Phaeophyta (brown algae), sex attractants of eggs of, 131–3
phagocytes, versatility of, 53–4
phenylalanine, *Dunaliella* and, 124
pheromones, of opposite mating types of slime moulds, 89, 106
phosphodiesterases: acting on cAMP, in slime-mould amoebae, 104–5, (mutants lacking), 98, 100; acting on cGMP, in fruiting *Myxococcus,* 163, 165
5-phosphoribosyl pyrophosphate, and fruiting in *Myxococcus,* 164
phosphotransferase transport system: enzymes II of, acting secondarily as chemoreceptors, 142–3
Physarum polycephalum (myxomycete), 115; calcium in control of motility of, 135; carbohydrates attracting not identical with those metabolized, 118–20; specificity of carbohydrate receptors in, 120
Phytophthora cinnamomi, 121, 122–3
Phytophthora infestans, 120
Phytophthora palmivora, attractants for zoospores of, 120, 121, 122–3
Phytophthora spp., chemotaxis of zoospores of, 117
pili, of *Myxococcus, see* fimbriae
pinocytosis, in polymorphonuclear leucocytes, 77, 78, 79
polarity, of locomoting leucocytes, 15, 31, 73
Polysphondylium pallidum (slime mould), 91, 101
Polysphondylium spp., 92–3, 98
Polysphondylium violaceum, 91, 101
procaine, neutrophils and, 36
prostaglandins, as chemotactic agents, 40, 42
proteases: released by neutrophils, 62, 64, 65; reversibly inhibit gliding motility in *Myxococcus,* 158
protein S, of *Myxococcus,* 165, 166
proteins: acting as transducers between receptors and tumble regulator in bacteria, 144–5, 148, 150; amphipathic and denatured, as chemotactic agents, 53, 54, 59–66; chemotaxis on surface coated with, 62–3, 64; methylated, at carboxyl of glutamate residue, in cell membrane, 105, 145–7, 150, (enzymes methylating and demethylating) 147–8, (multiplicity of forms of) 148–9; receptors for, on *E. coli,* 142, and on leucocytes, 61–2, 74
Protista, 116

protons, released in *Dictyostelium* at binding of cAMP, 103, 104
pseudopodia, of polymorphonuclear leucocytes, 73, 76
pteridine ring, in attractants for slime moulds, 89–91

raffinose, *Neocallimastix* zoospores and, 123, 124
random locomotion, 1; definitions of, 2
random walk, 2–5; relation of diffusion constant to, 8–9; in three dimensions, 139
receptors, *see* chemoreceptors
red blood cells, 68
relay system, in aggregation of slime-mould amoebae, 95–6, 98, 101
repellants: receptors for, in *E. coli,* 145; secreted by vegetative slime-mould amoebae, 93
rhapidosomes, cytoplasmic filaments below cell membrane in *Myxococcus,* 160–1
ribose: *E. coli* and, 142, 145; *Physarum* and, 118, 119
RNA polymerase of *Myxococcus,* 161, 164

Salmonella typhimurium, 139
Saprolegnia spp., 124
scalar fields (chemical concentration, intensity of illumination, substratum adhesiveness, temperature), 11
scalar quantity of environment, simple kinesis as response to, 16, 22–4
scatter diagrams, for testing for taxis or guidance, 17–19
serine, *E. coli* and, 143
serum albumin, chemokinetic agent, 40, 41, 42, 56, 59, 60; neutrophils and, 61–2, 64, 65
sex attractants, 127, 136; of *Allomyces,* 127–31; of brown algae, 131–3; of ciliated Protozoa, 134; for fern sperm, 29; mechanisms of response to, 134–5
shock reaction, 20
sirenin, sex attractant of *Allomyces,* 117, 127, 128, 129–30
slime moulds, cellular, xi, sii, 90; chemotaxis in: biochemical mechanism of, 101–5; during sexual phase, 106–7; in multicellular aggregate, 108–10; negative, in vegetative amoebae, 92–3, 95; towards food, 87–92; in wild-type and mutant, 100; *see also Dictyostelium, Polysphondylium*
sorbitol: *Neocallimastix* zoospores and, 123, 124
sorbose: *Neocallimastix* zoospores and, 123, 124; *Physarum* and, 118, 119
Staphylococcus aureus, 54
starvation: in aggregation of slime moulds, 89, 90, 95, 101; in fruiting of *Myxococcus,* 163–4
Stereum hirsutum (bracket fungus), *Badhamia* and, 118
Stigmatella aurantiaca (Myxobacterium), 156, 167
substratum: adhesiveness of, 10, 21; contact of leucocytes with, 28, 29, 33; curvature of, 16, 21
sucrose, *Physarum* and, 118, 119
superoxide radicals, produced by leucocytes, 44, 45
surfaces: fruiting of *Myxococcus* and, 163; leucocytes and, 45, 62–3, 64; swimming of *Ectocarpus* male gametes and, 133

3T3 cells, 4, 5, 6, 21
taxis, xii, 10–11; assay of, 17–19; in boundary responses, 19–22; coexistence of kinesis and, 10; as response to a vector quantity of the environment, 16, 22–4, (in three dimensions), 17; spatial mechanism of, 13–15; temporal mechanism of, 11–13; temporal and spatial, 23
temperature, leucocytes and, 33, 68, 77
tensor fields, 16–17; boundaries between isotropic fields as, 20–1
tensor quantity of environment, guidance as response to, 16, 22–4
β-2-thienyl alanine, stimulates fruiting in non-fruiting mutants of *Myxococcus,* 165
threonine, stimulates fruiting in *Myxococcus,* 165
tissue damage: attraction of leucocytes to sites of, 28; immune-complex-mediated, 55
topotaxis, 116
toxin causing immobilization of cells, appears as attractant, 116
transducers (proteins), connecting chemoreceptors to tumble regulator in bacteria, 144–5, 148, 150
tropisms, 29
tryptophan, *Dunaliella* and, 124
tumbling, in bacterial locomotion, 139–40; calcium, and balance between swimming and, 151; regulator of, acting on flagella, 144; transducers linking chemoreceptors to regulator of, 144–5, 148, 150
tyrosine, *Dunaliella* and, 124

vector fields (magnetic, gravitational, luminous), 11; boundaries as, 20
vector quantity of environment, taxis as response to, 16, 22–4

xylose, *Neocallimastix* zoospores and, 123, 124

zygote of *Dictyostelium,* cytophagic cell, 106–7